Microelectronic Systems
Level I

Units in this series

Microelectronic Systems

Level I

P. Cooke

TECHNICIAN EDUCATION COUNCIL
in association with
HUTCHINSON
London Melbourne Sydney Auckland Johannesburg

Hutchinson & Co. (Publishers) Ltd

An imprint of the Hutchinson Publishing Group

17–21 Conway Street, London W1P 6JD

Hutchinson Group (Australia) Pty Ltd
30–32 Cremorne Street, Richmond South, Victoria 3121
PO Box 151, Broadway, New South Wales 2007

Hutchinson Group (NZ) Ltd
32–34 View Road, PO Box 40–086, Glenfield, Auckland 10

Hutchinson Group (SA) (Pty) Ltd
PO Box 337, Bergvlei 2012, South Africa

First published 1982

© Technician Education Council 1982

Set in Times

Printed in Great Britain by The Anchor Press Ltd
and bound by Wm Brendon & Son Ltd
both of Tiptree, Essex

British Library Cataloguing in Publication Data
Microelectronic systems 1.
  1. Microelectronics
  I. Technician Education Council
  621.381′71   TK874

ISBN 0 09 147191 5

Contents

Preface

This book is one of a series on microelectronics/microprocessors published by Hutchinson on behalf of the Technician Education Council. The books in the series are designed for use with units associated with Technician Education Council programmes.

In June 1978 the United Kingdom Prime Minister expressed anxiety about the effect to be expected from the introduction of microprocessors on the pattern of employment in specific industries. From this stemmed an initiative through the Department of Industry and the National Enterprise Board to encourage the use and development of microprocessor technology.

An important aspect of such a development programme was seen as being the education and training of personnel for both the research, development and manufacture of microelectronics material and equipment, and the application of these in other industries. In 1979 a project was established by the Technician Education Council for the development of technician education programme units (a unit is a specification of the objectives to be attained by a student) and associated learning packages, this project being funded by the Department of Industry and managed on their behalf by the National Computing Centre Ltd.

TEC established a committee involving industry, both as producers and users of microelectronics, and educationists. In addition widespread consultations took place. Programme units were developed for technicians and technician engineers concerned with the design, manufacture and servicing aspects incorporating microelectronic devices. Five units were produced:

Microelectronic Systems	Level I
Microelectronic Systems	Level II
Microelectronic Systems	Level III
Microprocessor-based Systems	Level IV
Microprocessor-based Systems	Level V

Units were also produced for those technicians who required a general understanding of the range of applications of microelectronic devices and their potential:

Microprocessor Appreciation	Level III
Microprocessor Principles	Level IV

This phase was then followed by the development of the learning packages, involving three writing teams, the key people in these teams being:

Microelectronic Systems I, II, III — P. Cooke
Microprocessor-based Systems IV — A. Potton
Microprocessor-based Systems V — M. J. Morse
Microprocessor Appreciation III — G. Martin
Microprocessor Principles IV — G. Martin

The project director during the unit specification stage was N. Bonnett, assisted by R. Bertie. Mr Bonnett continued as consultant during the writing stage. The project manager was W. Bolton, assisted by K. Snape.

Self-learning

As an aid to self-learning, questions are included in every chapter. These appear at the end of the chapters with references in the margin of the chapter text (for example Q1.2), indicating the most appropriate position for self-learning use. Answers to each question are given at the back of the book.

The books in this series have therefore been developed for use in either the classroom teaching situation or for self-learning.

Introduction

This book is the first of a series aimed at students studying the design and application of microelectronic systems. The material covered in this book was written to the objectives specified in the TEC Unit Microelectronic Systems U79/602.

This book is therefore structured to cover the topics that must be understood to cover the study objectives. It has been specially written to act as a textbook for training courses that follow the TEC objectives. Additionally it has been written in a form suitable for self-study.

Since no previous knowledge of electronics or computers could be assumed the topics covered include a general introduction to both subjects. The slant is, however, very much towards microcomputers and microprocessor-based systems.

A particular feature of the book is that it provides an overview of microelectronics at the systems level. The idea is to provide ways of thinking, talking and writing about how microelectronic systems work, as well as explaining the basic ideas on which all computers rely. Hopefully this approach will make the study of the technical details more interesting and provide a firm foundation for further studies.

The last chapter is directed at helping you to write and run some real programs. To do this you must have access to a suitable computer. If at all possible you should be able to call on the assistance of a knowledgeable engineer or teacher.

As a guide to using the book it is suggested that you first quickly read Chapters 1 to 5, and perhaps even Chapter 6. Do not worry about understanding it all; that is a task for more careful study. However, this approach will help give you a better feel for the whole subject, and make the new words and ideas easier to cope with. Do not worry that the meaning of some words and phrases are not clear. If you understand them you would not need to study the book. To provide a basis for learning how to talk and think about microelectronics and its application is the book's primary aim.

Acknowledgements

Before writing this book I had written many papers for technical conferences and journals, but not a book. In other words I was used to writing for professional engineers. In consequence I am indebted to the forebearance of TEC staff, my wife and Peter Hill of Hutchinson, for trying to help me to write at a suitable level for an introductory textbook. I can only hope that you find the end product to be a suitable compromise between accuracy and simplicity.

Thanks must also go to Janice Parmenter for word-processing my handwritten script and to Dr Steven Harris of Cooke Associates for carefully reading and constructively criticising the material. The errors and omissions are, of course, mine.

PHIL COOKE

Chapter 1 Basic systems

Objectives of this chapter *When you have completed studying this chapter you should:*

1 Understand what a system is.
2 Be able to use diagrams to describe systems.
3 Know that systems need a source of energy.
4 Be able to describe some measurement transducers.
5 Understand what a controller is.
6 Be able to follow a simple flowchart.

1.1 Introduction to systems

This book is concerned with microelectronic systems. We will therefore first consider the question, what is a *system*? A dictionary definition is along the lines of *a collection of parts that work together so as to provide a complete function*. Some examples will help illustrate what is meant.

- Washing machine: a system for washing clothes.
- Oven: a system for cooking food.
- Power station: a system for generating electrical power.
- School: a system for educating children.
- Scales: a system for weighing things.
- Accounting: a system for keeping financial records.

Notice that there are all sorts of systems.

A diagram is often a very useful way of helping to describe a system. Figure 1.1 shows the general idea. Figure 1.2 shows the same idea in a very slightly different way. Notice that there are three important parts to the drawing. They are:

- *Inputs*
- *The process*
- *Outputs*

Inputs and outputs to a system take many forms, but they can usually be classified as:

- Physical things
- Energy
- Information

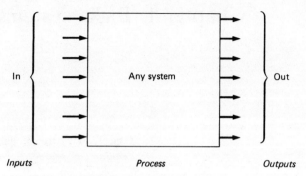

Figure 1.1   A very generalised system

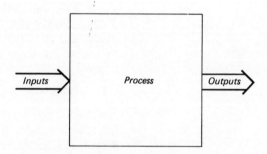

Figure 1.2   An alternative way of representing a system

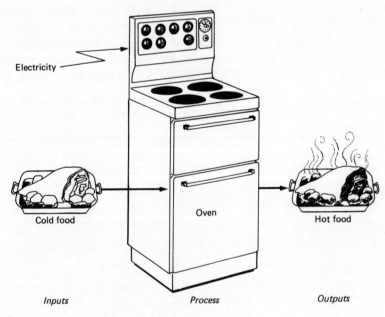

Figure 1.3   An electric oven viewed as a system

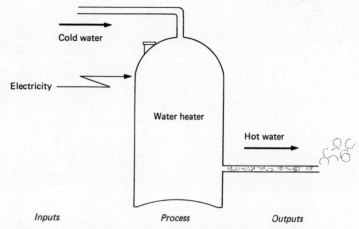

Figure 1.4 An electric immersion-heater-based hot water system

The process performed by the system is a much more individual thing. Microelectronic systems are primarily concerned with processing information.

If the meaning of the phrase *processing information* is not yet clear, the following system examples will help to clarify what is meant.

Consider first Figure 1.3, a diagrammatic representation of *an oven viewed as a system*. It explains, without the need for further words, that an oven is a system for converting cold food into hot food. It shows also that electricity must be input to the system. It is used as a source of energy. Another, and similar, example is an electric water heater. It is represented as a system in Figure 1.4. It features:

Inputs   • Cold water
           • Electricity
Outputs • Hot water

A slightly more complex example is an electronically controlled washing machine. As Figure 1.5 shows, there are now three inputs and two outputs. They are:

Inputs   • Cold water
           • Dirty clothes
           • Electricity
Outputs • Dirty water
           • Clean clothes

All the systems so far considered need a source of power. Obviously energy must be used if water or food is to be made hotter than it was. A supply of energy will also be needed in a radio. Here the miniscule amount of energy available at the aerial is amplified, i.e. magnified so as to produce the much larger power supplied to the loudspeaker. The extra energy must come from a power supply. Moreover some energy

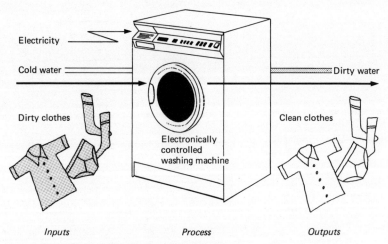

Electricity

Cold water

Dirty water

Dirty clothes

Electronically
controlled
washing machine

Clean clothes

Inputs   *Process*   *Outputs*

Figure 1.5   A washing machine as a system

will be wasted. In fact all electronic systems need a source of energy.
Without a power supply they *will not work*. Not only must the output
power be produced, but there must also be a supply of power even if
the inputs and outputs are purely informational.

Next consider a weighing machine. As is shown in Figure 1.6, this
features:

Inputs	●	Goods (to be weighed)
Outputs	●	Goods (weighed)
	●	Information (the weight)

Here the physical input and output are identical. However, the
system provides as an output the weight of the goods. In this case the
primary purpose of the system is to produce this piece of information.
So the process performed by a weighing system is to measure the
weight of the goods placed on it and, also, to produce as an output, a
number equal to the weight.

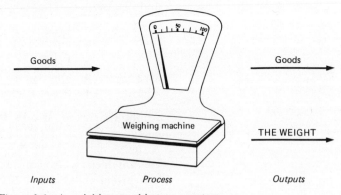

Goods   Goods

Weighing machine   THE WEIGHT

Inputs   *Process*   *Outputs*

Figure 1.6   A weighing machine as a system

As a final example consider an accounting system. It features:

Inputs • Records of financial transactions
Outputs • Statements of financial position

To achieve an accounting system the inputs are processed according to well defined rules. After various arithmetic processes, the required outputs are produced. In this example both inputs and outputs are information, i.e. *informational.*

It is because computers are very good at processing information that they are much used for helping keep financial records. However, any information that can be expressed as numbers, or letters, or lines, or images, can be processed by a computer. Thus a computer is a very general-purpose system for processing information. Moreover, by means of suitable input and output devices a computer can be used to measure and control. Consequently microcomputer-based systems are very versatile.

Q1.1, 1.2, 1.5

See note in Preface about questions

1.2 The use of diagrams

Diagrams have already been introduced as aids to describing systems. In this section we shall extend the range of concepts to which diagrammatic descriptions are used. Additionally we will consider some guidelines that will help you to produce drawings that are easy to understand. By referring back to the figures associated with Section 1.1 you should notice:

• *Lines:* they indicate the flow of information, or of things, or of energy.
• *Arrows:* showing the direction of flow.
• *Names:* explanations.
• *Boxes:* representing a process.

Notice how useful a box is. It can be used to represent virtually any process or piece of equipment. Notice also that we describe what is done and ignore how it is done. Thus a car engine can be described as a mechanism for converting petrol into rotary motion. When systems are described in this way we often refer to the arrangement as a *black-box.*

Microelectronic systems are made by connecting together microelectronic components. Each component has inputs and outputs. One particular output may be used to *drive,* i.e. to provide an input to one or more components. When the information is in electrical form, we talk of input and output *signals.*

Whilst a rectangular box can be used to represent any process it often occurs that a system contains many identical components. Then we often use a special and easily recognisable shape for frequently

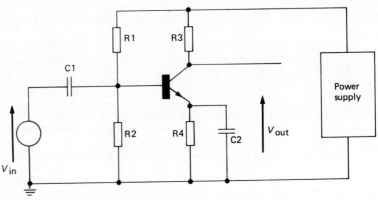

Figure 1.7   A simple electronic circuit

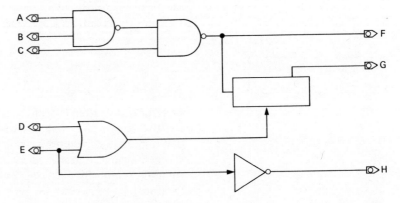

Figure 1.8   A simple logic circuit

Symbol	Representing
	Resistor
	Capacitor
	Transistor
	Voltage generator
	NAND gate
	OR gate
	Invertor

Figure 1.9   Symbols used in Figures 1.7 and 1.8

occurring units. Sometimes these *symbols* are agreed by international standards committees. By adhering to these recognised standards engineers can more easily understand manuals and drawings produced by other engineers. Circuit and logic diagrams, of which Figures 1.7 and 1.8 are examples, show some of the standard symbols used by electronics engineers. For interest, rather than because you need *at this stage* to be able to use them, the symbols are defined in Figure 1.9.

Next we will look at some of the ways in which diagrams are used to explain and define how a complicated process is achieved.

Figure 1.10 shows a record player. Such a diagram shows the general scheme of things and is often called a system *schematic*.

Another useful type of diagram shows *data flow*. An example is given in Figure 1.11 [1.1]. It shows the basic flow patterns in a system for a book publisher. Diagrams of this type are used by systems analysts when trying to describe how a particular system works.

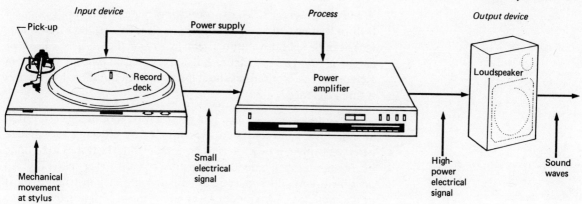

Figure 1.10 Schematic of a record player

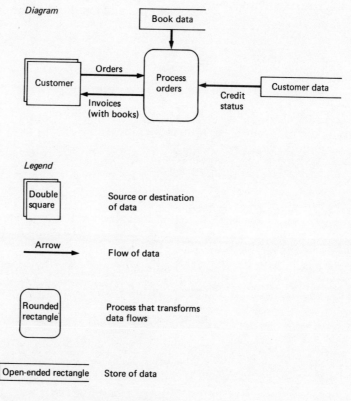

Figure 1.11 A data-flow diagram

When computers are used to process information the overall job must be broken down into a series of very simple tasks. To show how major tasks can be split down into a sequence of smaller tasks, we use *flow diagrams (flowcharts)*. Figure 1.12 shows an example, adapted from Reference 1.2. Notice that the steps have been numbered; this eases future reference.

When a process is described in a precise way, we call the description an *algorithm*. In everyday conversation we are often not very precise when we explain how a job is to be done. When the task is to be done by a computer each individual step must be defined very carefully. All possible eventualities must be anticipated and each instruction must be unambiguous. This is done by breaking the overall task into a large number of very simple tasks. The algorithm defines the order in which they are to be performed. When the first task is completed the second is started, and so on. Usually it will involve decision points. Then the process will vary according to the situation. For instance, as is shown in Figure 1.13, the tyre changing routine will need to be modified *if* the spare is found to be flat.

In the description of algorithms we frequently use the concept of *looping*. This technique allows an essentially repetitive task to be shown as a loop. For instance a computer that was programmed to

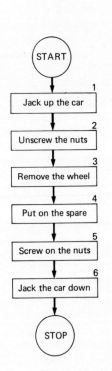

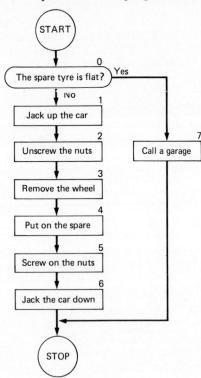

Figure 1.12　First flat tyre flowchart

Figure 1.13　Second flat tyre flowchart

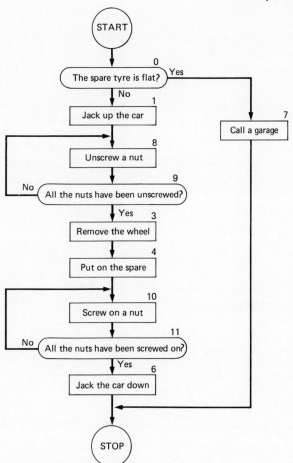

Figure 1.14 Final flat tyre flowchart

process a payroll would compute everyone's pay in the same way. After computing the pay for the first employee on its list, it would process the second, and third, and so on, until no more remained to be done. Similarly the actions associated with the wheel nuts in Figure 1.12 could be expressed as a loop, as is shown in Figure 1.14.

Finally we will look briefly at a *hierarchical* diagram. In one of its most familiar forms it is used to describe the way the management of a company is organised. Figure 1.15 shows an example. Figure 1.16 shows the way in which the control of an electric washing machine might be arranged. It shows the overall task broken down into three major sub-tasks, and hence into lesser tasks. This sort of diagram is more similar to the data-flow diagrams of Figure 1.11 than to the flowchart of Figure 1.13.

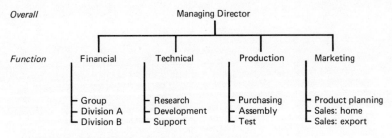

Figure 1.15   Hierarchy diagram for a company

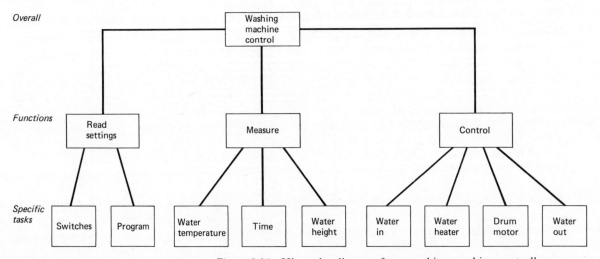

Figure 1.16   Hierarchy diagram for a washing machine controller

Q1.6, 1.7

Traditionally electronic systems are described with circuit diagrams and block schematics, a computer program with flowcharts. However, the hierarchical and data-flow type of diagram are increasingly used to describe how business systems, and electronic systems, and computer programs are organised.

1.3   Measurement transducers

Transducers are the heart of measuring systems. A dictionary definition is along the lines of *a device that receives information as energy in one physical form, and converts it into another.*

The mercury-in-glass thermometer is a familiar example. Temperature is converted into a column length. It is measured by reference to a scale, calibrated in degrees. They are very accurate because the mercury expands with temperature in a very predictable way.

To explain the way in which the output of a transducer is related to its input conditions we use a graph. Figure 1.17 shows the graph for a

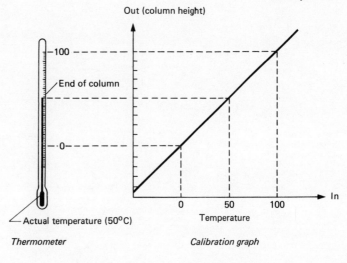

Thermometer   Calibration graph

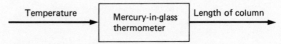

Figure 1.17 The mercury-in-glass thermometer

mercury-in-glass thermometer. To obtain it, temperature would be varied and the column height noted. Then, from the table of results, a graph would be plotted. In use the process is reversed. We then look at the output. Then from the graph or scale, we *deduce* what the input must be.

Measurement transducers suitable for use with microelectronic systems have electrical outputs. Fortunately there are large numbers of transducer types available. With them microelectronic systems can be provided with inputs that are measurements of, for example:

- Temperature
- Weight
- Speed
- Position
- Sound level
- Colour

In all cases the output is an electrical signal that gets bigger or smaller according to the input. By measuring the electrical output the input can be deduced by reference to the graph, or table, of input against output. In practice most transducers are arranged to have a graph that is a straight line; but this is not essential. Provided the input/ output relationship is always the same the electronics can be arranged to allow for graphs of almost any shape.

From a systems viewpoint a measurement transducer, suitable for

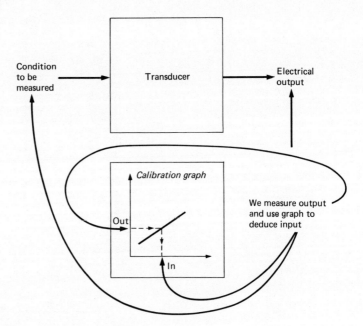

Figure 1.18   Systems view of a transducer

connection to a microelectronic system, can be represented in the way shown in Figure 1.18.

A *floating ball* is often used as the basis of a *liquid-level transducer*. The basic idea is used in the ball-cock used in WCs and cold water tanks. It is also used as the basis for petrol gauges in cars. The basic arrangement is shown in Figure 1.19. As the liquid level varies, the ball will float to different heights, causing the pointer to move. In effect the liquid level is being converted into an angular position.

To *calibrate* the arrangement for use as a transducer the tank would be emptied and the resultant indicator position marked *empty*. Similarly the tank would be filled and the *full* position marked. Intermediate levels would be calibrated in a similar way. The end result is a transducer with mechanical position as an output. Calibration allows us to see, directly, what the input is.

In cars the same basic arrangement is modified so as to give an electrical output. A float within the petrol tank is attached to a pivoted arm which, in turn, is used to move a slider along a resistance. The general arrangement is shown in Figure 1.20. It is then a simple matter to arrange that a meter varies its deflection according to the value of resistance and hence according to the petrol level. In fact, because it is so very easy to create an electrical signal that is a measurement of resistance, many transducers are based on resistors.

The resistance of platinum varies with temperature in a very predictable way. For platinum of high purity the graph of resistance

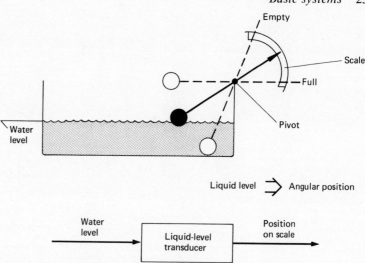

Figure 1.19 A float-based level transducer

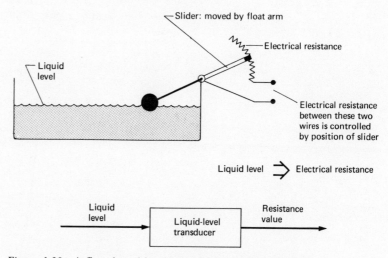

Figure 1.20 A float-based level transducer – electrical output

against temperature is so predictable that all *platinum resistance thermometers (PRTs)* follow the same graph. So instead of having to calibrate each PRT over the whole temperature range, it is sufficient to standardise them at just one temperature. The same graph can then be used for all PRTs.

Usually it is arranged that the electrical resistance is exactly 100 Ω at 0°C. Then at 100°C the resistance will be 139 Ω. At 50°C, which is half-way between 0°C and 100°C, the resistance will be almost exactly half-way between 100 and 139 Ω, i.e. 119.5 Ω. So by measuring the resistance, temperature can be deduced and therefore measured. A

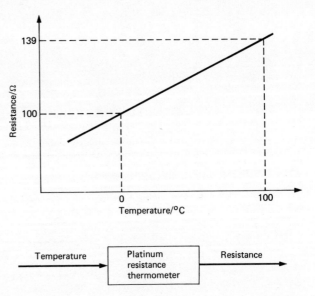

Figure 1.21   Input/output relationship for a platinum resistance thermometer

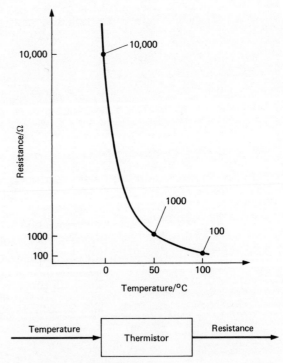

Figure 1.22   Input/output relationship for a typical thermistor. (Note: logarithmic scale on vertical axis.)

graph of resistance against temperature for a platinum-resistance-thermometer is shown in Figure 1.21.

Thermistors are another form of resistance-based temperature transducer. They are very cheap. Instead of being made from a pure metal, they are made from composite material. Their virtue is that a small change in temperature causes a very large change in resistance. The rate at which resistance changes with temperature is much higher than is the case with a platinum-resistance thermometer. This helps to keep the associated electronics cheap. However, the graph of resistance against temperature is not a straight line. A typical example is shown in Figure 1.22. When, as here, the graph is not a straight line the transducer is said to be *non-linear*. Microprocessors can be programmed to allow for this.

In this section we have so far studied just a few of the transducers commonly connected to microelectronic systems. Other examples include:

- *Pick-up on a record player* The output is an electrical signal that *follows* the movement of the stylus.
- *Touch-switch* The output depends on whether or not the switch is currently being touched.
- *Silicon light-sensor* This device gives an output current that is a measure of light intensity. It is used in automatic cameras.
- **Q1.4, 1.9, 1.10, 1.14, 1.21** *Tachometer* The output is a measure of rotational speed.

1.4 Output mechanisms

To convert the electrical output signals of microelectronic systems into other physical forms, a variety of output mechanisms are used. They are the output equivalent of input transducers, but have electrical inputs instead of outputs. Examples are:

- *Loudspeaker* To convert electrical signals into sound waves.
- *Lamp* To convert electricity into light.
- *Cathode ray tube (CRT)* The output device used in a television to convert electrical signals into pictures.
- *Thyristor* A special semiconductor component. It allows a low-level electrical signal to control a large electrical load.
- *Motor* A device to convert electrical energy into rotary mechanical motion.
- *Valve* Electrically actuated valves allow electrical signals to open and close valves, so controlling liquid flows.
- *Segmented display* The seven-segment display allows the presence and absence of seven electrical signals to be used to create an illuminated representation of the numbers 0,1,2,3,4,5,6, 7, 8, 9. Layout of the segments is shown in Figure 1.23.
- *Printer* Some electric typewriters, and printers used with

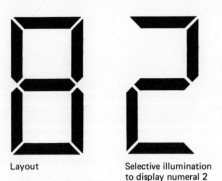

Layout   Selective illumination
                          to display numeral 2

Figure 1.23   The seven-segment display

computers, contain a mechanism whereby low-level electrical signals can cause the movement of paper and the printing of symbols (numbers and letters).

In this section we are not concerned with how output mechanisms work. Instead it is sufficient to be able to think of output arrangements as mechanisms for converting from an electrical signal into the **Q1.11**   desired form.

1.5   Everyday systems

In this section we will examine several systems found in everyday life. This will provide an opportunity for bringing together the ideas used in the earlier sections. We will also introduce the concept of *feedback* control. This is a very general principle of major importance to any self-regulating system.

A thermostat is the most familiar example of an electrical feedback mechanism. The general idea is that the system is arranged to do whatever is necessary to make the temperature equal to that pointed to by the control knob. We use them to keep living and working areas at a fairly constant temperature, regardless of the weather. They are also used to regulate the internal temperature of refrigerators and deep freezers. They are also used to regulate hot water temperature. Additionally they are used in highly sophisticated form within industry. In fact the list of applications is almost endless.

In general a feedback system is used to automatically regulate something, so that it is *made to be* equal to the desired value. We call this the *set-point*. Feedback systems can be used to regulate, for example:

● Temperature
● Position
● Speed
● Humidity

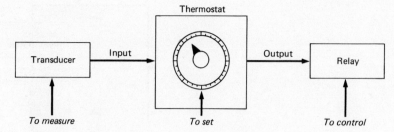

Figure 1.24 A thermostat is used to automatically control temperature

The way in which a thermostat works can be better understood by reference to Figure 1.24. It features,

- *Input:* from a transducer, which measures temperature.
- *Knob:* this enables the user to determine the set-point.
- *Output:* in this case a *relay* is used so that a large electrical load can be switched on and off.

The system algorithm is as follows:

- Measure actual temperature.
- Compute difference between set-point and actual.
- If temperature too low, switch on heat.
- If temperature too high, switch off heat.
- Keep repeating algorithm for ever.

Based on the data-flow diagrams and flowchart methods introduced in Section 1.2, the system can also be described in the ways shown in Figures 1.25 and 1.26.

Before leaving the example of a thermostat it is important to notice that the system has an input that represents the *user's instructions*. Some of the systems diagrams used in Section 1.1 were difficient in

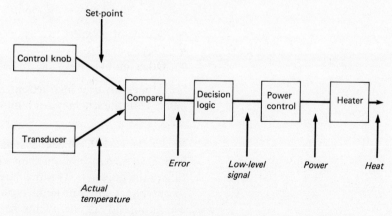

Figure 1.25 Thermostatically controlled heater as a data-flow diagram

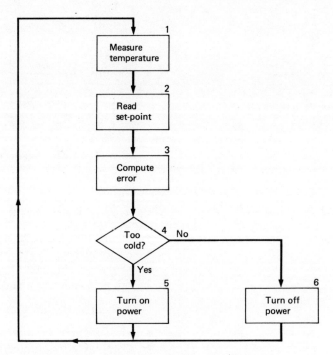

Figure 1.26   Flowchart for a thermostatic controller

this respect. In particular Figures 1.3, 1.4 and 1.5 omitted *user instructions*. A thermostatically controlled water heater provides for the desired water temperature to be set via a control knob. The electric oven must be similarly set to the desired temperature. There may also be other controls, such as switch-on time via a built-in clock. From a systems point of view an on/off switch also provides user instructions. The washing machine needs even more complex instructions. So, in general, a microelectronic system is best viewed as having the following three classes of inputs:

- Source of power
- User instructions
- Other inputs

Sometimes the user instructions are, in effect, built into the equipment. Thus the schematic of Figure 1.27 is an improvement on Figure 1.2.

Before leaving feedback systems it is interesting to note that biological systems contain many examples. For instance the temperature of the human body is regulated, by an internal feedback system, to an almost constant temperature. The *set-point is built in at birth*. The conversion of energy into heat is automatically controlled so as to keep the body temperature constant.

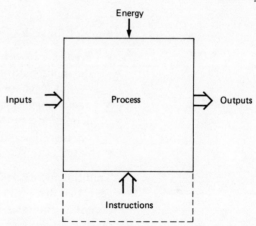

Figure 1.27 Many systems can be represented in this form. Note that some or all of the instructions may be built in

A radio is our next example of a system. It features:

- *Main input:* radio signals from aerial.
- *User inputs:* via knobs and switches – to select station, adjust volume or turn on/off.
- *Energy input:* electricity.
- *Output:* sound-waves via loudspeaker.

It should now be clear that most organisations and all manner of equipment can be viewed as a system with inputs and outputs. Micro-electronic systems are a particular class of system in which both inputs and outputs are low-level electrical signals. Suitable input and output arrangements are added to produce microprocessor-*based* equipment. This permits:

- Inputs and outputs to be in convenient form.
- Complex relationships between inputs and outputs.

Q1.12, 1.13, 1.22 In the next section we shall examine the sort of processes that microcomputers can be programmed to carry out.

1.6 Controllers

In the preceding sections we have seen how a system is a mechanism that generates outputs by processing the inputs according to a well defined algorithm. The algorithm is built in to a microelectronic system, but can usually be modified, or alternatives selected, by means of user inputs. In many systems, and especially in micro-processor-based systems, the processing system performs the function of a *controller*.

In microelectronic systems the inputs and outputs are low-level electrical signals and in controllers the sort of tasks done are one or more of the following:

- *Feedback control*   This usually involves a transducer to measure the input conditions, a user input of the desired set-point and an output mechanism.
- *Sequencing*   This usually involves a clock relative to which actions are timed, inputs of actuator positions and outputs to on/off valves and switches.
- *Displays* to show what is happening.

Traditionally electric sequencers and controllers involve motors and gears that drive cams and spring-loaded contact sets. Time clocks are everyday examples of this type of arrangement. Electric washing machines use more complex mechanisms of a similar type.

Microprocessor-based controllers record time in a way that is now popularised by the digital watch. They are based on specially designed microelectronic circuits that work in the same sort of way as microprocessors, but are much simpler. In both cases time is measured by counting the vibrations of an electrically stimulated quartz crystal oscillator. This is in keeping with the time-honoured clockmaker's principle: a mechanism with a very constant period is made to drive a counting arrangement. The microelectronic clock is very accurate because the rate at which the quartz crystal vibrates is very stable. It is not much affected by temperature or age.

When controllers are based on microprocessors there is a control algorithm that defines the process. How it causes the inputs to control the outputs may be very complex. However the microprocessor is arranged so that it is very easy for design engineers to change and define the algorithm. When the algorithm is in a form acceptable to the microprocessor it is called a *program*.

In microprocessor-based equipment the program is usually built in. It is installed when the equipment is manufactured and will not be destroyed when the equipment is switched off. The program is then said to be in *firmware*. This is in contrast to computer programs entered via keyboards or from magnetic disks or tapes. Then the program can be easily changed or accidently corrupted and is said to be *software*.

Before considering an example of a controller, the concept of *internal condition or state* needs to be defined. It arises because computers have memories. Thus future actions can be arranged so that they are determined not only by current inputs, but also by internal states. For instance, a switch can be made to have no effect unless an internal, i.e. built in, clock shows the time to be in a particular range, say between 8.00 a.m. and 10 a.m. Another possibility is a system in which an output is generated only if an input occurs after a particular event has

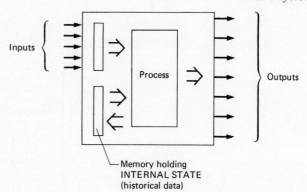

Figure 1.28 Process inputs are current inputs *plus* internal state

been previously detected. Thus the earlier ideas, whereby the outputs were controlled only by the current inputs needs to be modified so *the outputs are controlled according to the inputs and the internal state.*

The idea of internal state, as an extension to the current inputs, is shown in the schematic of Figure 1.28. In the internal state area it is possible to remember information for as long as desired. Digital watches do this. It always keeps a record of what time it is. Similarly a record could be kept of how long a particular output or input had been on, or how many times a particular event had been detected. Because the memory circuits used to store such data are very cheap, some systems are made to remember a lot of historical data.

Before proceeding further it will be useful if we summarise and consolidate our view of a microelectronic controller. Firstly it is a system with information as both inputs and outputs in the form of low-level electrical signals. The way in which the input data (past and present) is processed so as to produce the outputs is defined by an algorithm. A microcontroller is in fact a special-purpose computer, and the algorithm is in a form that a computer can obey, i.e. a computer program. By changing the program we can change the way in which the outputs are controlled by the inputs. Later in Chapter 4 we shall examine how computers work. At present it is sufficient to know that a computer program is an algorithm in a form that a computer can follow.

Finally in this section we will examine the sort of algorithm that would be built into a microelectronic controller for a washing machine.

Q1.8

In Figure 1.29 is shown a schematic of an electric washing machine. Each of the important sub-systems are numbered, for ease of reference. Their functions are as follows:

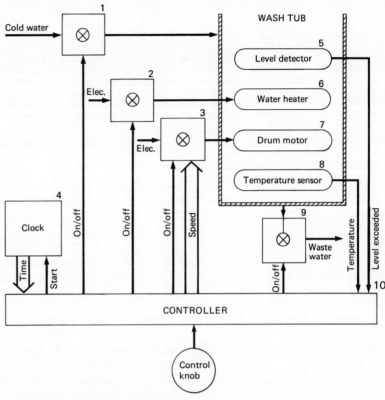

Figure 1.29 Block schematic of a washing machine

Block 1

Name   Cold water valve
Inputs   Cold water (physical)
            On/off signal (informational)
Outputs   Cold water
Function An electrically operated valve to enable a low-level
            on/off electrical signal to control the water supply.

Block 2

Name   Heater controller
Inputs   Electricity
            On/off signal (informational)
Outputs   Electricity
Function An electrically operated switch to enable a low-level
            on/off electrical signal to control the water heater.

Block 3

Name   Motor controller
Inputs   Electricity
            On/off signal (informational)
            Speed data (informational)

Outputs   Electricity
Function   A unit to control the supply of electricity to the drum motor so that its speed can be set to the required value.

Block 4
Name   Clock
Inputs   Crystal oscillator (built in)
          Stop/start signal (informational)
Outputs   Elapsed time (informational)
Function   An electric stop-watch with electronic control and read-out facilities.

Block 5
Name   Water-level detector.
Inputs   Water-level (measurement/informational)
Outputs   Level reached signal (informational)
Function   A measurement transducer that outputs a signal indicating if the water level has reached, or exceeded, the pre-set level.

Block 6
Name   Water heater
Inputs   Electricity (energy)
Outputs   Heat (energy)
Function   The electric immersion heater is used to convert electricity into heat, and thereby increase water temperature.

Block 7
Name   Drum motor
Inputs   Electricity (energy)
Outputs   Rotary motion (energy)
Function   The motor is used to rotate the drum. The drum is a multipurpose component. During the washing cycle the drum is used as an agitator. It is also used, at higher speed, as a spin dryer.

Block 8
Name   Temperature sensor
Output   Electrical signal (informational)
Function   A temperature transducer to provide an electrical signal that is a measure of the water temperature.

Block 9
Name   Dump valve
Inputs   Waste water (physical)
          On/off signal (informational)
Outputs   Waste water (physical)
Function   An electrically operated valve used to enable dirty water to be dumped.

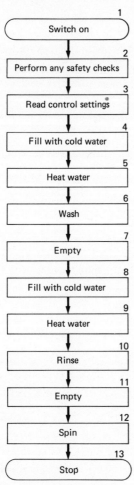

1
Switch on

2
Perform any safety checks

3
Read control settings

4
Fill with cold water

5
Heat water

6
Wash

7
Empty

8
Fill with cold water

9
Heat water

10
Rinse

11
Empty

12
Spin

13
Stop

Figure 1.30 Flowchart for washing machine controller

Block 10

Name	Controller
Inputs	User program (informational)
	Clock (4) (informational)
	Water level (5) (informational)
	Temperature (8) (informational)
Outputs	Control signal for (1) (informational)
	Control signal for (2) (informational)
	Control signal for (3) (informational)
	Speed signal for (3) (informational)
	Control signal for (9) (informational)
	Control signal for (10) (informational)
Function	The controller, a microelectronic unit, deduces the washing programs requested by the user, by reading the positions of the control knobs and switches. Then, in accordance with the programs stored in firmware, it manipulates the output control signals, using the input signals as appropriate.

To complete the description of the system, the process performed by the controller is described in Figure 1.30. It is a first level flowchart; it leaves quite a few questions unanswered but gives a good overview. It is typical of the flowchart that we use when designing or first explaining a computer program.

To complete the description of the algorithm followed by the controller each of the flowchart steps 2 to 12 inclusive must be expanded. Thus each of these steps is described by a more detailed flowchart. Several examples are shown in Figures 1.31, 1.32 and 1.33. These flowcharts include several features of interest:

1 All three include a loop. In all cases the idea is the same; a particular condition is sought, and until it is detected no further action takes place.

2 As Figures 1.32 and 1.33 show, several of the main flowchart steps can be described by identical sub-flowcharts. In both examples the technique used is the same. Early in the algorithms the *required conditions* are determined. Thus when heating water for washing, the water temperature required is obtained by reference to a stored table of values for the selected wash program. When preparing rinse water the required rinse water temperature is similarly obtained. In like manner the appropriate time and speed is obtained in the *rotate drum* steps.

3 Notice that in Figure 1.33 several sequential and closely related tasks are grouped together. This helps to make a flowchart more easily understood.

Step 4   Fill with cold water

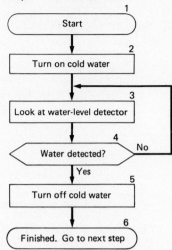

Figure 1.31 Flowchart for Step 4 of washing machine control flowchart

Steps 5 & 9   Heat water

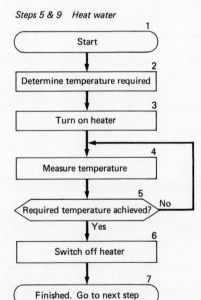

Figure 1.32 Flowchart for Steps 5 and 9 of washing machine control flowchart

Steps 6, 10 & 12   Rotate drum

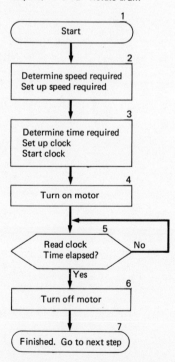

Figure 1.33 Flowchart for Steps 6, 10 and 12 of washing machine control flowchart

Q1.3–1.20, 1.23–1.25

4   Note that the controller we have been studying is a micro-electronic device. Built into it is a program that causes the pre-defined sequence of events.

References

1.1   Forsythe, Keenan, Organick and Stenberg, *Computer Science, a First Course*, pp4–8, John Wiley and Sons, New York (1977).

1.2   C. Gane and T. Sarson, *Structured Systems Analysis: Tools and Techniques*, p9, Prentice Hall, New Jersey (1979).

Questions

1.1   Can a system have more than one input?

1.2   Can a system have many outputs?

1.3   In what form is information presented to a microelectronic system?

1.4   Why are transducers used to generate input signals to microelectronic systems?

1.5   Why do microelectronic systems need a power supply?

1.6   How is the process performed by a system described?

1.7   What is an algorithm?

1.8   What inputs are used by the process algorithm?

1.9   What do transducers do?

1.10   Make a list of transducers that can be used as input devices.

1.11   List some output transducers suitable for use with electronic systems.

1.12   List some systems.

1.13   For each of your answers to Questions 1.11 and 1.12, list the inputs and outputs.

1.14   Describe how the following measurement transducers work:
(a)   a resistance-based thermometer;
(b)   a float-based level detector.

1.15   Refer to Figure 1.29 which shows the schematic for a washing machine. Without reference to later figures explain how it works.

1.16   Draw a flowchart for Steps 7 and 11 of the flowchart of Figure 1.30.

1.17   Are there any problems in implementing the flowchart of the previous question?

1.18   With the help of Figures 1.29–1.33, produce a hierarchical type of diagram (see Figure 1.16) for the control program.

1.19 What is the purpose of a flowchart?

1.20 What is the primary purpose of inputs and outputs to a micro-electronic system?

1.21 What do we expect to be able to deduce by looking, i.e. measuring, the output of a sensor?

1.22 What is a program?

1.23 What do systems do?

1.24 What is a controller?

1.25 What determines the algorithm followed by a controller?

Chapter 2 Analogue and digital systems

Objectives of this chapter *When you have completed studying this chapter you should:*

1   *Understand the difference between analogue and digital.*
2   *Be able to give examples of analogue quantities.*
3   *Know that computers are digital systems.*
4   *Understand a truth table.*
5   *Understand the idea of binary information.*
6   *Know the difference between parallel and serial.*

2.1   Analogue and digital quantities

Most physical measurements are continuous quantities. They are called *analogue* quantities. Examples include:

- Weight
- Temperature
- Length
- Volume
- Speed

In all cases the quantities are such that they can change by an imperceptible amount. We measure them by comparison with *standards*. Length, for instance, is measured by comparison with a calibrated ruler.

Digital quantities are naturally expressed as whole numbers. There is a minimum amount by which digital numbers can change.

A quantity of ball bearings is a digital quantity. In contrast the length of a piece of string is an analogue quantity. However, when an analogue quantity is measured, it has to be expressed as a number, which is a digital quantity.

A particularly important type of digital quantity occurs when there can only be one of two values. Then it is called a *binary* quantity. Thus an electric light switch is a binary device. It is either *on*, or it is *off*. Except for very brief periods when it is changing rapidly from one state to the other it is either *on* or *off*.

It so happens that electronic circuits which have binary input and output signals are cheap to produce. Moreover, systems of unlimited complexity can be made by using more and more of these basic circuits.

By convention, electronic systems made from binary circuits are called *digital systems*. Digital computers are a particular, and very important, class of digital system.

Within a digital system all the signals are binary signals. By convention we call these *digital signals*. Thus digital signals are always one of two values. By convention we use the symbols 0 (zero) and 1 (one). We could use any other pair of symbols, but to avoid confusion we stick to 0 and 1. In later sections we shall see how all types of information can be represented by digital signals. At present it is sufficient to accept that *microprocessors* and other electronic digital computers are all based on digital signals. They are made by connecting together large numbers of digital circuits.

Analogue quantities can also be represented by electrical signals. Examples of this type were discussed in Section 1.3. The resistance thermometer led to an electrical output signal that varied in sympathy with the temperature experienced by the transducer. The floating-ball petrol gauge led to an electrical signal that varied, and was a measure of, the quantity of petrol in the tank. Another type of transducer is used in the pick-up of a record player. The mechanical movement of the stylus, as it follows the record's groove, is converted into an output signal. An *amplified*, i.e. magnified, replica of this signal is used to drive a loudspeaker. The loudspeaker creates sound waves, which we hear. In this example three analogue quantities are involved:

- The shape of the record groove.
- The electrical signal.
- The sound waves.

A clue to the distinctive feature of this type of analogue signal lies in the words *shape* and *wave*. They correctly imply something that changes with time. In fact, most electronic signals of an analogue type change very rapidly with time. Radio signals change on time scales that are measured in nanoseconds (10^{-9} s). This is approximately equal to the time it takes light to travel 30 cm. Even in audio electronics the signals can oscillate (change backwards and forwards) at rates up to 20,000 times per second (20 kHz).

Audio signals, and the electrical signals that represent them, have graphs of signal amplitude against time of the type shown in Figure 2.1. Because graphs of this type are classified by their shape, we talk of the signals having a *waveform*.

The purpose of this introductory section is to make clear the features that distinguish analogue and digital systems. Later sections will

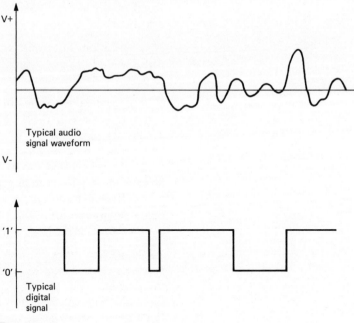

V+

Typical audio
signal waveform

V-

'1'

'0'

Typical
digital
signal

Figure 2.1

concentrate on one system or the other, and the digital type in particular. However, before proceeding further it is important that you can easily distinguish one type of system or quantity from another. When in any doubt it may help to focus attention on the nature of the informational signals. Analogue quantities change smoothly and continuously. This is not to say they cannot change rapidly, but from one instant to the next, the change is almost imperceptible. Analogue signals are therefore said to be *continuous*. In contrast digital signals are *discontinuous*. Reverting back to the example of an electrical switch, the fundamental difference in nature of analogue and digital quantities should be clear.

Q2.1–2.3

2.2   Analogue-based systems

Electronic systems made by connecting together electronic circuits that manipulate electrical signals of an analogue type are called *analogue systems*. The individual circuits have analogue input and output signals and are referred to as *analogue circuits*. Well known examples of analogue systems include radio, TV and hi-fi. Additionally, science and industry use many types of analogue measurement system.

Virtually all early electronics was of an analogue type. Initially digital circuits were not used. Now an ever-increasing fraction of all electronic circuits are digital. Moreover many systems include circuits of both types. Thus it is increasingly difficult to find examples of pure analogue systems. Furthermore, many computer-based systems are connected to analogue subsystems at inputs and outputs. The remainder of this section will be devoted to outlining the way in which analogue circuits are used in a few well known electronic systems.

First we will consider the telephone system. A simplified schematic is shown in Figure 2.2. The system is required to perform two basic functions. A user of a telephone connected to exchange A requires that a *connection* is made to, let us say, a particular telephone in exchange B. This is a digital function. The user wants to be connected to no-one (phone not in use), or someone in particular. When the telephones are *connected* a speech signal must be sent and received. Speech signals are analogue. The handset microphone converts the sound waves into analogue electrical signals for transmission. At the receiving hand-set a small transducer converts the electrical signals back into sound waves. So, in a telephone system we have a digital routeing network, defined by numerical dialling codes. In addition we have an analogue transmission path.

Our next example is a radio receiver. It receives, via the aerial, an input signal from all the radio stations in the world that are powerful enough to be detectable by that aerial. To select a particular station the tuning knob, which is an analogue control, is turned to select the wanted transmitter. Everything about this part of the system is analogue. Volume and tone controls are also of an analogue type. But the switches are digital controls.

Finally we will examine a system that is less well known, but heavily used by industry for monitoring conditions at a remote point. Companies with North Sea oil platforms, for instance, sometimes have highly sophisticated control and monitoring rooms on the mainland. The principles involved are the same as those used by space scientists when monitoring and controlling a space capsule.

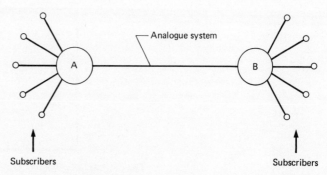

Figure 2.2 A simplified telephone network

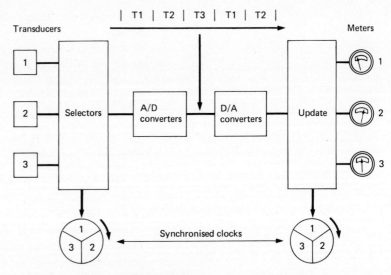

Figure 2.3   Remote monitoring system

Figure 2.3 shows a very much simplified schematic that isolates the basic principles involved. The fundamental idea is that each of the three meters is made to display the analogue quantity measured by one of the three transducers. Instead of three signal paths, there is just one. For very long distances radio is used; for shorter distances just one pair of wires is required.

Information is transmitted in digital form in a way explained in the flow chart of Figure 2.4. The new system elements introduced are *converters*. An analogue-to-digital converter, often abbreviated to *A to D* or *A/D*, is an electronic circuit that produces a digital output that is a measure of the analogue input. In other words it produces a

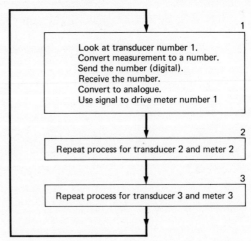

Figure 2.4   Flowchart describing the operation of a remote monitoring system

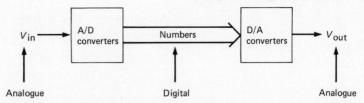

Figure 2.5 A system with analogue in and analogue out, but with digital transmission

number that is a measure of the analogue input signal. A digital-to-analogue converter, sometimes called a *D to A* or *D/A* performs the reverse process. Thus a system of the form shown in Figure 2.5 is analogue in and analogue out. Nevertheless the information is transmitted digitally, as a number.

To complete our study of Figure 2.3 notice that the information is transmitted by continuously scanning the input transducers. Thus, the information is sent sequentially. Looking at the transmission path we would see first, T1, then T2 then T3, then T1, then T2, etc. By having a synchronised clock the information can be unscrambled. When information is sent in this way we say it is *time-multiplexed*. We:

- read and send
- wait a while
- read and send again
- and so on.

Provided we send a measurement so frequently that it has hardly changed before we measure it again, we have lost nothing of significance. The technique is important because the waiting time for one channel can be used to send data for many other channels. Whilst detailed study of systems of this sophistication are beyond the scope of this book the example is important because it highlights three important features of microelectronic systems:

- *Converters* Circuits of this type enable conversion between analogue and digital signals.
- *Operating speeds* Microelectronic circuits measure and process information so fast that they can often be arranged to do many jobs in sequence, yet appear to be doing everything continuously.
- *Time-multiplexing* The technique whereby one signal transmission path is made to do the jobs of several (otherwise parallel arrangements) is much used in microprocessor-based systems.

Q2.9–2.14

2.3 Digital circuits

In earlier sections a digital circuit was defined as having input and output signals that are binary in nature. In most microelectronic systems this is arranged in the following way:

Binary 1 is represented by +5 V.
Binary 0 is represented by 0 V.

It would make equal sense to reverse the definitions, but unless everyone uses, and sticks to, a consistant definition there is much confusion. So, by convention, the most positive of the two signal levels is called a 'one', *unless stated otherwise*.

Digital circuits perform logic functions. They are therefore usually referred to as *logic circuits*. Whilst it is outside the scope of this book to explain the detail of logic system design, it is important in understanding digital systems to have some appreciation of what logic circuits do. So we will examine the basic functions of AND, NOT and NAND. The latter is an abbreviation of NOT AND. First we will look at a logic circuit that performs the AND function – shown diagrammatically in Figure 2.6. It features:

- *Inputs:* A and B.
- *Outputs:* C.
- *Process:* see truth table in Figure 2.6.

A *truth table* shows all the possible combination of inputs.

With two inputs there are only 2 × 2 = 4 possible combinations of input signals. For each of the possible combinations the output signal is shown. Thus the output is uniquely defined by the inputs. Moreover by looking at the output we can deduce certain things about the input signals. In particular, if, and only if, both A *and* B is a 1, is the output a 1. For this reason this type of logic circuit is called an AND gate. The name *gate* arises as a consequence of simple logic circuits having certain similarities to a gate. Like everyday gates they can be open and closed. Electronically they can be used to either enable or block an input signal from affecting the output.

A NOT function merely makes the output the opposite of the input. It is defined in Figure 2.7. For obvious reasons this circuit is usually called an *invertor*.

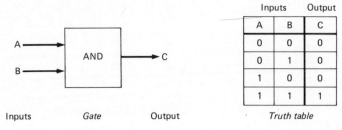

Inputs		Output
A	B	C
0	0	0
0	1	0
1	0	0
1	1	1

Inputs   Gate   Output   *Truth table*

Figure 2.6   The AND gate

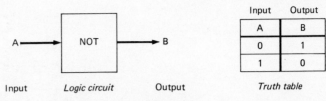

| Input | Output |
A	B
0	1
1	0

Input   *Logic circuit*   Output   *Truth table*

Figure 2.7 The invertor or NOT function

Finally we will look at the NAND function. It can be made by connecting together an AND gate and an invertor. It is, by convention, allocated a special symbol of the shape shown in Figure 2.8. Before leaving this example you may like to study its truth table. Did you notice that the output is a 1 if *either* of the input signals is a 0? This observation may help to emphasise that such circuits produce an output that is logically related to the inputs.

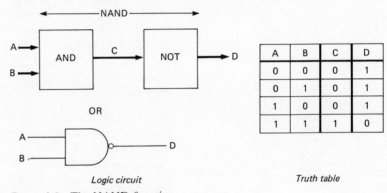

A	B	C	D
0	0	0	1
0	1	0	1
1	0	0	1
1	1	1	0

Logic circuit   *Truth table*

Figure 2.8 The NAND function

Next we will consider the way in which logic signals change. In fact, as was explained in earlier sections, they change rapidly from one state to the other. Nevertheless, short as this time may be, it is important to avoid using signals that are changing. If at any time this restriction is broken, the result is a malfunction of the equipment.

A typical logic signal is shown in Figure 2.9. It depicts a signal that occasionally changes state; it also identifies those periods of time when the signal must not be used.

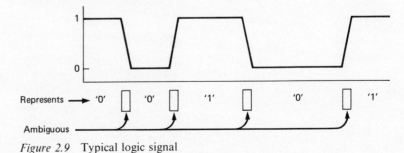

Figure 2.9 Typical logic signal

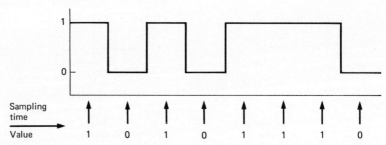

Figure 2.10   Time is usually divided into equal periods

To avoid using logic signals at times when they are *liable* to be changing, designers take care to use the signals only when they are assured that they will be steady. This is done by introducing special timing signals. They define the periods when the signals will be steady. Figure 2.10 shows how eight regularly spaced *sampling* periods result in a waveform being interpreted as the binary sequence 10101110. This is another example of time-multiplexing. Over eight units of time, eight bits of data have been transmitted. Note the word *bit*. It is used to describe a single piece of binary information. Thus each bit may be a 1 or a 0.

When information is transmitted bit by bit, sequentially over one transmission path, the data is said to be *serially* transmitted. This is an important concept which Figure 2.11 may help to consolidate. It shows eight people each holding one bit of data, filing in sequence through a door. They present the information serially.

In contrast Figure 2.12 shows the same data being presented in *parallel*. In this case the information can be seen at a glance. In the serial mode it takes time and a good memory to acquire the data. In microelectronic system terms the serial mode of data transmission is slow but cheap. The parallel mode needs more wires and costs more but is faster.

Next we will examine one of the most important of all logic circuits; the *memory*. In its simplest form it is a device that can be set to either 1 or 0. It remembers forever the state into which it was last set. Its state is available as an output signal. The function is summarised in Figure 2.13.

Individual memory circuits are often referred to as latches.

In the early computers, i.e. in the 1950s, memory circuits were very expensive, difficult to design, slow to operate, bulky, unreliable and consumed so much energy they were too hot to touch. In 1981 over 64,000 single bit memory cells are routinely made on a *single chip* of silicon a few millimetres square. They are very fast, very reliable, consume little power and cost very little.

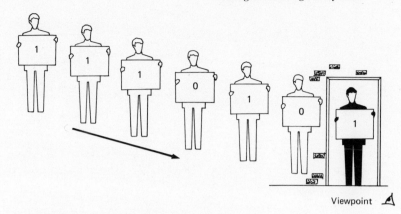

Figure 2.11 Data being presented serially

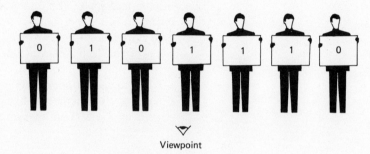

Figure 2.12 Data presented in parallel

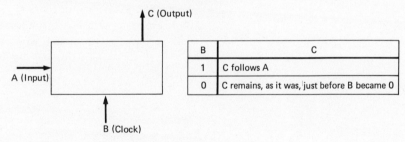

Figure 2.13 The memory circuit

Before proceeding to look at how logic circuits and signals are used we will briefly examine three logic functions that are much used in digital electronics. In all cases they can be made from the simple gates already described. They are all used as components for micro-computer systems.

The decoder

This circuit has as many outputs as there are possible *combinations* of the inputs. With two inputs there are $2 \times 2 = 4$ outputs, with three inputs there are $2 \times 2 \times 2 = 8$ outputs, and so on. Figure 2.14 shows a

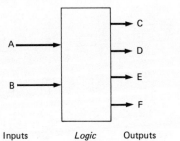

A	B	C	D	E	F
0	0	1	0	0	0
0	1	0	1	0	0
1	0	0	0	1	0
1	1	0	0	0	1

Inputs   *Logic*   Outputs   *Truth table*

Figure 2.14   The two-input decoder

two-input and four-output decoder, together with a truth table. Its notable feature is that for each of the four possible combinations of inputs, *just one* of the four outputs is a 1. The outputs could, for example, be made to control four lights. If a 1 caused *light-on* and a zero caused *light-off*, there would always be just one light that was on. So, in effect, each of the possible input combinations, or *codes*, selects just one output. Decoder circuits are used within the memory chips used in microcomputers. By manipulating just 10 input wires, any one out of 1024 memory cells can be selected. Because the input code defines and selects an internal *location*, it is called an *address*.

Selection switches

To enable one of two or more sources of information to be selected there are logic circuits of a suitable type. By studying Figure 2.15 you will see that they are similar to decoders in the sense that the address acts as a *selector*.

Data transfer

When information stored in one place is wanted in another, the data must be transferred. The principle is straightforward, as a study of

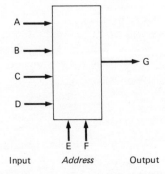

E	F	G
0	0	Same as A
0	1	Same as B
1	0	Same as C
1	1	Same as D

Input   *Address*   Output   *Truth table*

Figure 2.15   The digital switch

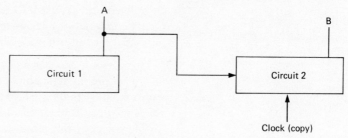

Figure 2.16 Data transfer (1 to 2)

Figure 2.16 should reveal. It features two of the basic memory units described first in Figure 2.11. When the *clock* terminal of Circuit 2 is pulsed to a 1, the output of B is made equal to the output of the memory, Circuit 1. Thus the *contents* of Circuit 1 have been *copied* to Circuit 2.

Q2.5–2.8

By convention the copying operation described is referred to as *loading* or *moving*. It is important to notice that moving does not destroy, or in any way alter, the original, i.e. *source*, information.

2.4 Digital representation of information

In the previous section it was shown, by means of a truth table, that two bits of binary information could occur in only $2 \times 2 = 4$ different ways. Three bits of binary data can occur in $2 \times 2 \times 2 = 8$ different ways. In general n bits can be combined in 2^n different ways. The significance of this relationship will be better appreciated by studying Table 2.1.

Table 2.1

Number of bits	Number of combinations
1	2
2	4
3	8
4	16
5	32
6	64
7	128
8	256
9	512
10	1024
11	2048
12	4096
13	8192
14	16,384
15	32,768
16	65,536

By convention it is usual to refer to 1024 as 1K, *when* talking about binary combinations. Using this convention 16,384 becomes 16K. In particular 65,536 becomes 64K (64,000 × 1.024).

In the previous section we saw how an address was used to select one out of all the possible combinations. Thus a 10-bit address is used with a memory that has 1K of memory locations. Similarly to address a 64K memory we need 16 address lines. Most microcomputers cater for memories of this size.

The idea of using a binary number as an address is fundamental to all digital computers. It enables the selection of just one out of 2^n addresses. To reinforce this concept the mechanism is pictorially described in Figure 2.17. It shows how the address acts as a *pointer*.

When a binary number contains n bits it is said to be an *n-bit number*. Thus binary 10 is a 2-bit number, whereas 10101 is a 5-bit number.

Within a computer system a *word* is a collection of bits. The number of bits making up a word is standardised for a particular machine. In some computers it may be 8 bits, in others 16 bits and in some it may even be 64 bits. But no matter how wide or narrow the standard size, it is defined as a *word*.

In principle the word size can be any convenient number, but by common agreement it is usual to keep to a multiple of 8. Words or units of 8 bits are called *bytes*. Most second-generation micro-processors have a word size of 8 bits, i.e. one byte wide. More advanced microcomputers usually have a word of 16 bits, i.e. two bytes wide.

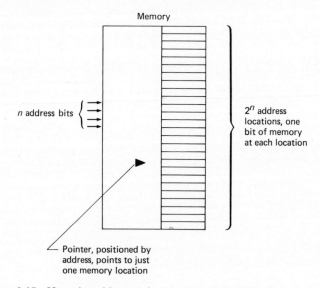

Figure 2.17   How the address points to the selected location

Now we will shift our attention to the way in which binary data is used to represent information. The smallest unit is a single binary bit. It can be represented by a square containing a 1 or a 0. Often such a single bit is used to indicate the status of some situation. It could, for instance, be used to indicate if a particular event had, or had not, occurred. Then the bit of data is called a *flag* or *status bit*.

Next we will examine the way in which binary words can be used to represent decimal numbers. The idea can be easily seen by using a truth table type of approach, as is shown in Table 2.2. Notice that there is a *one-to-one* correspondence. A particular code represents a particular number. For obvious reasons we refer to the binary data as a *binary number*.

Table 2.2 How decimal numbers can be represented in binary

Binary number	Decimal number
0000	0
0001	1
0010	2
0011	3
0100	4
0101	5
0110	6
0111	7
1000	8
1001	9
1010	10
1011	11 etc.

Whilst there are many other possible schemes of associating numbers and binary combinations, the one given has many advantages. In particular it leads to a straightforward set of rules for adding and subtracting. In consequence the relationships shown are an industry standard.

Besides using binary words to represent numbers and addresses, we must and do use many other coding schemes. We will briefly consider the two most important ones.

Alphanumerics

To communicate between people we usually either speak or write. In the latter case we use standardised symbols. In the Western world we use two sets:

- The alphabet
- Numbers

In total this seems to involve $26 + 10 = 36$ symbols. However we also use punctuation marks. Also the letters may be capitals or smaller ones (upper or lower case). Even so 7 bits can represent 128 different symbols. This is more than enough. In fact there is an internationally agreed allocation for all the necessary symbols. It is called the American Standard Code for Information Interchange and is invariably referred to as *ASCII Code* (from its initial letters). Some of the allocations are shown in Table 2.3. (A complete table is given in Appendix 3.)

Table 2.3   Some ASCII code allocations

Binary code	Letter
1000001	A
1000010	B
1000011	C
1000100	D

Finally we will look at the way binary data is used to identify and define computer programs. This requires a little anticipation of the way in which computers work; a detailed explanation is given in Chapter 4. At present it is sufficient to know that a computer obeys a repetitive and cyclical action of the sort shown in Figure 2.18.

Having introduced the idea that a computer is a device that obeys one instruction after another, we will look at the way binary words are used to represent the instructions. Once again it is merely a matter of associating a number with a name, in this case the name of the instruction that is represented. Exactly *how* the computer can perform the action is described in later chapters. At present we are only concerned with seeing how a sequence of binary words can describe an algorithm. It is therefore necessary to believe that a fairly

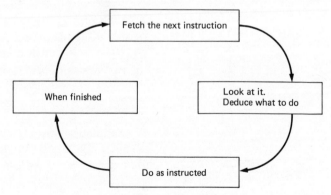

Figure 2.18   The cyclical beat of a computer

small number of basic operations is sufficient to enable the precise description of algorithms of any complexity we can imagine.

In summary a computer can perform a number of fairly simple and well defined tasks. Each of the different tasks is called an *instruction*. Typical of the functions performed by computer instructions are:

- Load
- Add
- Subtract
- Compare
- Wait
- Loop

Q2.4, 2.16–2.30

To represent these instructions the allocation of Table 2.4 could be used. In practice a computer has more instructions of great subtlety, but the concept is identical to that described. In both cases the action to be performed can be deduced by looking at the instruction code.

Table 2.4 How binary words are used to represent computer instructions

Binary code	Instruction
00000000	Load
00000001	Add
00000010	Subtract
00000011	Compare
00000100	Wait
00000101	Loop

Questions

2.1 Are digital signals continuous?

2.2 Is there any limit to the number/levels or states that an analogue signal can have?

2.3 Make a list of analogue quantities.

2.4 Why are converters used?

2.5 How many different states has a binary signal?

2.6 What electrical signal is usually used to represent a binary 1?

2.7 Do microelectronic logic circuits need a supply of energy?

2.8 Why are digital circuits often called logic circuits?

2.9 Why are amplifiers used?

2.10 In what type of systems are amplifiers used?

2.11   Why do we talk of signals having a waveform?

2.12   Are switches an analogue device?

2.13   In what way is a radio transmission similar to a wire?

2.14   Why is time-multiplexing used?

2.15   Draw and complete a truth table for a three-input decoder.

2.16   How many rows of combinations would there be in a truth table for a logic circuit with eight inputs?

2.17   How many bits are there in a byte?

2.18   How many bits are needed for a word of two bytes?

2.19   Why are logic signals sampled?

2.20   What is the difference between serial and parallel transmission?

2.21   What is the function of a memory address?

2.22   How many storage locations are there in a memory with 10 address bits?

2.23   What is meant by *loading a memory*?

2.24   Does the move operation destroy information at the source?

2.25   What is a flag?

2.26   What is the ASCII Code?

2.27   After a computer has finished obeying the last instruction, what does it do?

2.28   How are computer instructions represented?

2.29   List examples of systems that are:
(a)   digital;
(b)   analogue.

2.30   Are resistance-based temperature transducers digital devices?

Chapter 3 Microelectronic components

Objectives of this chapter *When you have completed studying this chapter you should:*

1 Know that the hardware is the physical part of an electronic system.
2 Know the difference between discrete components and integrated circuits.
3 Know that integrated circuits consist of a large number of microscopic circuit elements.
4 Know that SSI, MSI, LSI and VLSI indicate increasing levels of integrated circuit complexity.

3.1 Discrete-component electronic circuits

When electronic circuits are made by connecting together *individual* components they are said to be *discrete-component* circuits. This distinguishes them from microelectronic circuits, i.e. those in which the complete circuit is made within a single piece of silicon. The individual circuit elements cannot be separated.

All electronic circuits are made by connecting together individual circuit *elements*. Each *physically separate* unit we call a *component*. So, in a discrete-component circuit each component is just one circuit element. The elements are mainly resistors, capacitors and transistors. They are described below.

Resistors

Resistors are the most common of all electronic components. They resist the flow of electricity to an extent that is given by their resistance value. The unit of measurement is the ohm. In abbreviated form it is indicated by the symbol Ω (the Greek capital omega).

Resistors are often of large values and the symbol 'k' is used to indicate thousands of. Thus a resistance of 1 kΩ = 1000 Ω.

Resistors are made from metals or other composite materials. Discrete-component resistors are usually a few millimetres in diameter, about one centimetre long and have a connecting wire at each end. Their resistance value is shown, not by a printed number, but by coloured rings. Each colour represents a number according to an internationally agreed and followed colour code.

Symbol	Represents

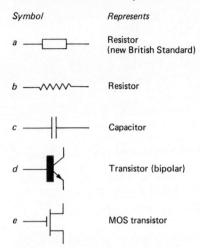

a — Resistor (new British Standard)

b — Resistor

c — Capacitor

d — Transistor (bipolar)

e — MOS transistor

Figure 3.1 Symbols used in circuit diagrams

When resistors are shown in circuit diagrams they are represented by the symbol shown in Figure 3.1, line *a*. American-based organisations and design engineers still tend to use the symbol shown in Figure 3.1, line *b*. When labelling circuit diagrams, resistors are referred to by the letter R. Thus a circuit with four resistors would use the labels R1, R2, R3 and R4.

Capacitors

Capacitors store electrical energy and, like resistors, have two wires. They can also be used to resist the *change* in electrical conditions. The extent to which they do this is given by their capacitance. It is measured in farads (symbol F). However, this is an impractically large unit and real capacitors have values that are measured in picofarads (pF), nanofarads (nF) or microfarads (μF).

Capacitors are bigger and much more expensive than resistors. They are used less.

When capacitors are shown in circuit diagrams they are symbolically represented in the way shown in Figure 3.1, line *c*. When labelling circuit diagrams capacitors are referred to by the letter C. Thus, in a circuit with four capacitors, they would be called C1, C2, C3 and C4.

Because capacitors resist changes in electrical conditions, they are often used to try and *prevent* voltages changing. Then we use a value which is sufficiently large, i.e. big enough. In electronic circuits capacitors used in this way are called *decoupling* capacitors. Often they are connected across power supplies, near to the circuit. When used in this way it is useful to think of a capacitor as a small battery, that charges to whatever voltage it is connected to. Once charged it tries to maintain its voltage.

Inductors

Inductors, like resistors and capacitors, are a basic circuit element. They are made from coils of wire and are therefore bulky. Consequently, they are used as little as possible, and avoided almost completely in microelectronic circuits. We shall not refer to them further.

Transistors

Transistors are called *active* components because they can amplify. Resistors and capacitors cannot amplify and are therefore called *passive* components.

Transistors are made from very pure silicon (not to be confused with silicone). The processes involved in making transistors are very sophisticated and have taken many years to develop. On a single slice of silicon a large number of identical transistors are made. Then the slice is broken into small rectangular dice, or chips. Each silicon chip is a few millimetres square.

The individual chips contain a complete transistor but to use it three wires must be connected. For ease of handling the device is mounted in a can, or encapsulated in plastic. From the package emerge three wires to connect the transistor into the circuit. Transistors packaged in this way are usually less than 1 cm in diameter and 1 cm long.

The symbol used to represent a transistor is shown in Figure 3.1, line *d*. Applying a small electrical signal to the input terminal causes a much larger change at the output terminal. In analogue circuits the circuit arrangements are such that the input and output waveforms are the same shape.

In digital circuits the input signal is used to make the transistor look like a switch. In one condition it looks like a very small resistance, in the other like a very large resistance.

MOSTs

MOS transistors (MOSTs) differ from the transistors described in the last section. They use much less power and are also much smaller. To distinguish the two types of transistors we call the original type *bipolar* and the newer type MOS. Exactly why the names are used is not easy to explain, and does not concern us.

To represent the MOS device the symbol shown in Figure 3.1, line *e*, is used. There are two varieties; the P-channel and the N-channel. For brevity they are called NMOS and PMOS. Both work in the same way, and have very similar properties. One is easier to make but the other is faster. The most obvious difference between them is that the direction of the current flows depend on whether the device is P- or N-channel. One is the complement of the other.

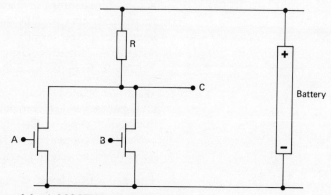

Figure 3.2   A MOST-based logic circuit

Some microelectronic circuits use P- *and* N-channel MOS transistors. They are called Complementary MOST circuits and are normally referred to as CMOS types.

MOSTs are not much used as discrete components. They are only cheap when made very small. Then they are extremely slow unless they are connected together by extremely short wires. MOSTs are thus used mainly within microelectronic circuits. Figure 3.2 shows a simple MOS logic circuit.

Printed circuits

Electronic circuits are now almost always made by the use of a printed circuit. Instead of connecting together the components by ordinary pieces of wire a wiring pattern is made by a photographic process (photolithography). The method of manufacture is briefly described below.

Printed circuits are made from special insulating board. On one side there is a thin layer of copper. The components are placed on the other side. For each component lead a hole is drilled through the board, and the leads are fed through the holes. On the copper side each lead is soldered to the copper. With non-processed board this would connect together all the wires from all the components. Instead the board is first processed so as to remove most of the copper, leaving only connections, or tracks, between the wires which are to be connected together. This is achieved by a photographic process.

The wiring pattern is defined by a drawing (the artwork). Black is used to show where copper is to remain; white indicates that copper is to be removed. From the master drawing a full size photographic film is made. Areas where copper is to remain are translucent, the other areas black.

Printed circuits are made by first coating the copper with a thin layer of a special light-sensitive chemical called photo-resist. The photographic film defining the connection pattern is then laid on top of the sensitised copper, and the sandwich is exposed to a strong light. Thus light reaches the sensitised layer only where copper is to remain. After development the exposed areas of photo-resist are very hard. Areas not exposed to the light are left unchanged.

When the printed circuit is subsequently immersed in suitable acids all the unexposed layer of photo-resist is eaten away together with the underlying copper. The hardened layer resists the etchants and protects the copper below. Thus there remains a pattern of copper as defined by the original artwork.

The protecting layer of hardened photo-resist is removed by a suitable solvent to leave a clean and precise pattern of conducting

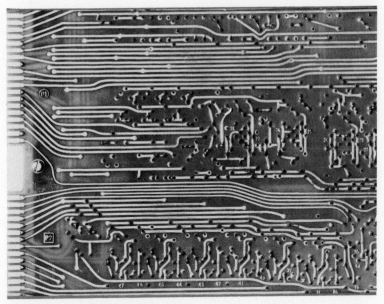

Figure 3.3   A printed circuit

tracks. A photograph of a printed circuit is shown in Figure 3.3. The manufacturing process is summarised in Figure 3.4.

Virtually all electronic equipment is made by mounting components on printed circuits. This is because it is cheap and easy to make large quantities of circuits to the same wiring pattern. It is a well proven method for mass-producing low-cost units.

With printed circuits each discrete component is mounted on the surface. Normally the wiring pattern requires more surface area than is required by the components. This is because each track must be a certain width and must be spaced from others. The result is that one component usually requires several square centimetres of surface area. So, in practice, a printed circuit of about 150 mm x 100 mm would be required for a circuit containing about forty discrete components.

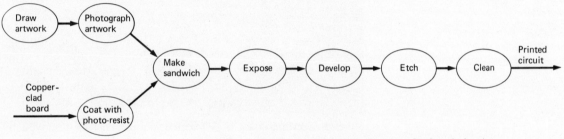

Figure 3.4   Process used for manufacture of printed circuits

Q3.1, 3.5, 3.7–3.13

The reason for describing the printed circuit manufacturing process at some length is that microelectronic circuits are also made by methods based on photolithography. The processes are therefore similar, but are much more elaborate.

3.2　Integrated circuits

Integrated circuits are complete circuits made within a single piece of silicon. For brevity they are often called *ICs*. Like printed circuits their manufacture involves the use of artwork and photographic processes. By the use of many processes all the different circuit elements can be made and interconnected within a small area of silicon. Each circuit element is arranged to be electrically separate despite all of them being *monolithic*, i.e. all within a single silicon chip.

Integrated circuits are made by a more elaborate version of the processes used to manufacture transistors. Many identical circuits are made in a single slice of silicon. Each rectangular chip contains an identical circuit. The dimensions involved are so small that the chips must be viewed through a microscope. The circuit has a number of

Figure 3.5　A silicon chip (much enlarged)

layers, each the result of photographic exposure, etching and processing. The outermost layer is a pattern of interconnections. Lower layers define transistors, resistors and capacitors. Figure 3.5 shows a magnified view of a silicon chip.

Each chip is a complete functional circuit. It cannot be broken down into circuit elements. The elements are integrated into a single chip component that is, on its own, a complete functional circuit. They are ideal building blocks for making electronic systems. To distinguish them from discrete-component circuits they are called integrated or microelectronic circuits. The popular press calls them silicon chips.

In summary, an integrated circuit is a complete functional unit made from a large number of microscopic circuit elements, i.e. transistors, resistors and capacitors, all within a single silicon chip.

A complete circuit can have many inputs and many outputs. This compares to a transistor which has only three connections. To connect the chip is not easy for it is very small. For this reason the chip is encapsulated in a rectangular package from which six to forty or more wires emerge. Each 'wire' is in fact a piece of flat metal, a *pin*. For ease of use a range of standard packages have been internationally agreed to. Collectively they are referred to as *Dual In-Line*, or *DIL* packages. An example is shown in Figure 3.6.

Since each DIL contains a functional circuit it is described by a functional diagram, i.e. a diagram showing how the inputs control the outputs. This contrasts with a circuit diagram, such as that of Figure 1.7.

Figure 3.6 A dual in-line package

54/74 FAMILIES OF COMPATIBLE TTL CIRCUITS

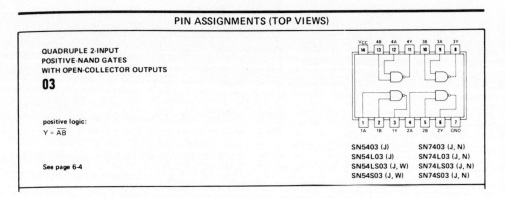

PIN ASSIGNMENTS (TOP VIEWS)

QUADRUPLE 2-INPUT
POSITIVE-NAND GATES
WITH OPEN-COLLECTOR OUTPUTS

03

positive logic:
Y = $\overline{AB}$

See page 6-4

SN5403 (J)   SN7403 (J, N)
SN54L03 (J)   SN74L03 (J, N)
SN54LS03 (J, W)   SN74LS03 (J, N)
SN54S03 (J, W)   SN74S03 (J, N)

Figure 3.7 Extract from data sheet for the SN7403 NAND gate package *(Courtesy: Texas Instruments)*

In manufacturers' data sheets are shown functional diagrams and other data required by a designer. Figure 3.7 shows the pin assignments for an SN7403 made by Texas Instruments. It is a circuit containing four identical circuits of the NAND variety. From this diagram we can deduce how the outputs are affected by the inputs. We can also see what each pin is connected to, and where to connect the power supply. Design engineers call these diagrams pinouts.

The manufacturer's data sheet for the SN7403 also shows the data reproduced as Figures 3.8 and 3.9. Whilst understanding this level of detail is not at present important, these extracts give some idea of what to expect in a manufacturer's data sheet. This level of detail may also act as a warning. It requires a lot of knowledge and experience to design or to use integrated circuits.

Initially integrated circuits were often thought to be important, only because they were very small. Now we know better. They are used mainly because they are very cheap and reliable. Moreover, because design engineers can put together systems without getting involved in circuit design, they can quickly realise very complex systems. They use very carefully designed integrated circuits. These are designed by specialist engineers and then mass-produced at low cost.

In the mid-1960s when integrated circuits were first available, the circuits were very simple. Anything more complex than a low-performance amplifier of a single logic gate was too complex to make. Even so, a discrete-component circuit that previously required perhaps 100 cm^2 of printed circuit, could be made in a single package of less than 1 cm^2. Suddenly electronic equipment became much smaller.

POSITIVE-NAND GATES AND INVERTERS WITH OPEN-COLLECTOR OUTPUTS

recommended operating conditions

Parameter		SERIES 54 / SERIES 74 ('01, '03, '05, '12, '22)			SERIES 54H / SERIES 74H ('H01, 'H05, 'H22)			SERIES 54L / SERIES 74L ('L01, 'L03)			SERIES 54LS / SERIES 74LS ('LS01, 'LS03, 'LS05, 'LS12, 'LS22)			SERIES 54S / SERIES 74S ('S03, 'S05, 'S22)			UNIT
		MIN	NOM	MAX	MIN	NOM	MAX	MIN	NOM	MAX	MIN	NOM	MAX	MIN	NOM	MAX	
Supply voltage, V_{CC}	54 Family	4.5	5	5.5	4.5	5	5.5	4.5	5	5.5	4.5	5	5.5	4.5	5	5.5	V
	74 Family	4.75	5	5.25	4.75	5	5.25	4.75	5	5.25	4.75	5	5.25	4.75	5	5.25	
High-level output voltage, V_{OH}	54 Family			5.5			5.5			5.5			5.5			5.5	V
	74 Family			5.5			5.5			5.5			5.5			5.5	
Low-level output current, I_{OL}	54 Family			16			20			2			4			20	mA
	74 Family			16			20			3.6			8			20	
Operating free-air temperature, T_A	54 Family	-55		125	-55		125	-55		125	-55		125	-55		125	°C
	74 Family	0		70	0		70	0		70	0		70	0		70	

electrical characteristics over recommended operating free-air temperature range (unless otherwise noted)

PARAMETER	TEST FIGURE	TEST CONDITIONS†	SERIES 54 / SERIES 74 ('01, '03, '05, '12, '22)			SERIES 54H / SERIES 74H ('H01, 'H05, 'H22)			SERIES 54L / SERIES 74L ('L01, 'L03)			SERIES 54LS / SERIES 74LS ('LS01, 'LS03, 'LS05, 'LS12, 'LS22)			SERIES 54S / SERIES 74S ('S03, 'S05, 'S22)			UNIT
			MIN	TYP‡	MAX	MIN	TYP‡	MAX	MIN	TYP‡	MAX	MIN	TYP‡	MAX	MIN	TYP‡	MAX	
V_{IH} High-level input voltage	1, 2		2			2			2			2			2			V
V_{IL} Low-level input voltage	1, 2	54 Family			0.8			0.8			0.6			0.7			0.8	V
		74 Family			0.8			0.8			0.6			0.8			0.8	
V_{IK} Input clamp voltage	3	V_{CC} = MIN, I_I = §			-1.5			-1.5						-1.5			-1.2	V
I_{OH} High-level output current	1	V_{CC} = MIN, $V_{IL} = V_{IL}$ max, V_{OH} = 5.5 V			250			250			50			100			250	µA
V_{OL} Low-level output voltage	2	54 Family V_{CC} = MIN, I_{OL} = MAX		0.2	0.4		0.2	0.4		0.15	0.3		0.25	0.4			0.5	V
		74 Family V_{CC} = MIN, V_{IH} = 2 V		0.2	0.4		0.2	0.4		0.2	0.4		0.35	0.5			0.5	
		Series 74LS I_{OL} = 4 mA											0.25	0.4				
I_I Input current at maximum input voltage	4	V_{CC} = MAX, V_I = 5.5 V			1			1			0.1			0.1			1	mA
		V_I = 7 V																
I_{IH} High-level input current	4	V_{CC} = MAX, V_{IH} = 2.4 V			40			50			10			20			50	µA
		V_{IH} = 2.7 V																
I_{IL} Low-level input current	5	V_{CC} = MAX, V_{IL} = 0.4 V / 0.3 V / 0.5 V			-1.6			-2			-0.18			-0.4			-2	mA
I_{CC} Supply current	7	V_{CC} = MAX			See table on next page													mA

† For conditions shown as MIN or MAX, use the appropriate value specified under recommended operating conditions.
‡ All typical values are at V_{CC} = 5 V, T_A = 25°C.
§ I_I = -12 mA for SN54'/SN74', -8 mA for SN54H'/SN74H', and -18 mA for SN54LS'/SN74LS' and SN54S'/SN74S'.

Figure 3.8 Part of data sheet for NAND gates and inverters *(Courtesy: Texas Instruments)*

POSITIVE-NAND GATES AND INVERTERS WITH OPEN-COLLECTOR OUTPUTS

schematics (each gate)

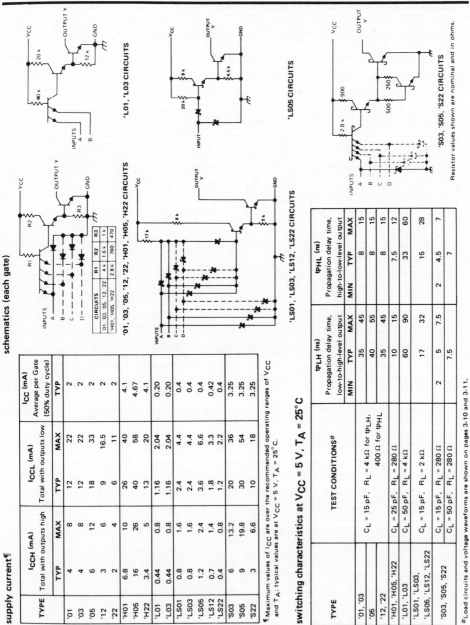

'L01, 'L03 CIRCUITS

'LS05 CIRCUITS

'S03, 'S05, 'S22 CIRCUITS

Resistor values shown are nominal and in ohms.

'01, '03, '05, '12, '22, 'H01, 'H05, 'H22 CIRCUITS

'LS01, 'LS03, 'LS12, 'LS22 CIRCUITS

CIRCUITS	R1	R2	R3
'01, '03, '05, '12, '22	4 k	1.6 k	1 k
'H01, 'H05, 'H22	2.8 k	760	470

supply current¶

TYPE	I_{CCH} (mA) Total with outputs high		I_{CCL} (mA) Total with outputs low		I_{CC} (mA) Average per Gate (50% duty cycle)
	TYP	MAX	TYP	MAX	TYP
'01	4	8	12	22	2
'03	4	8	12	22	2
'05	6	12	18	33	2
'12	3	6	9	16.5	2
'22	2	4	6	11	2
'H01	6.8	10	26	40	4.1
'H05	16	26	40	58	4.67
'H22	3.4	5	13	20	4.1
'L01	0.44	0.8	1.16	2.04	0.20
'L03	0.44	0.8	1.16	2.04	0.20
'LS01	0.8	1.6	2.4	4.4	0.4
'LS03	0.8	1.6	2.4	4.4	0.4
'LS05	1.2	2.4	3.6	6.6	0.4
'LS12	0.7	1.4	1.8	3.3	0.42
'LS22	0.4	0.8	1.2	2.2	0.4
'S03	6	13.2	20	36	3.25
'S05	9	19.8	30	54	3.25
'S22	3	6.6	10	18	3.25

¶Maximum values of I_{CC} are over the recommended operating ranges of V_{CC} and T_A; typical values are at V_{CC} = 5 V, T_A = 25°C.

switching characteristics at V_{CC} = 5 V, T_A = 25°C

TYPE	TEST CONDITIONS#	t_{PLH} (ns) Propagation delay time, low-to-high-level output			t_{PHL} (ns) Propagation delay time, high-to-low-level output		
		MIN	TYP	MAX	MIN	TYP	MAX
'01, '03	C_L = 15 pF, R_L = 4 kΩ for t_{PLH}, 400 Ω for t_{PHL}		35	45		8	15
'05			40	55		8	15
'12, '22			35	45		8	15
'H01, 'H05, 'H22	C_L = 25 pF, R_L = 280 Ω		10	15		7.5	12
'L01, 'L03	C_L = 50 pF, R_L = 4 kΩ		60	90		33	60
'LS01, 'LS03, 'LS05, 'LS12, 'LS22	C_L = 15 pF, R_L = 2 kΩ		17	32		15	28
'S03, 'S05, 'S22	C_L = 15 pF, R_L = 280 Ω	2	5	7.5	2	4.5	7
	C_L = 50 pF, R_L = 280 Ω		7.5			7	

#Load circuits and voltage waveforms are shown on pages 3-10 and 3-11.

Figure 3.9 Continuation of Figure 3.8

Exactly how many circuit elements can be put into a silicon chip depends on a number of things. In practice it depends critically on two main factors. The bigger the chip that can be made, the more it can hold. The smaller the area required by each component the more that can be accommodated. It is the latter of these two factors that dominates. In turn this is decided by the minimum dimensions that can be used, e.g. how narrow the wiring pattern can be.

Due to technological developments the semiconductor industry has, for many years, continually managed to put more and more circuitry onto a chip. When integrated circuits were first made it was with difficulty that the simplest possible circuits were made. First-generation integrated circuits had about ten circuit elements. By 1981 circuits with over 100,000 components had been made.

Gordon Moore, one of the engineers who played a key role in the development of silicon transistors and integrated circuits, predicted in the 1960s that each year the achievable complexity would double. For over fifteen years it has. Obviously improvements will not forever continue at this pace, but progress is expected until at least the late 1980s.

As we have seen, the first generation of integrated circuits contained just one functional circuit. Within a few years many circuits could be made on one chip. An example of this level of complexity was shown in Figure 3.7. Four identical circuits are contained in a single package with 14 pins. Eventually the number of simple circuits that could be put onto a single chip required more pins than could be reasonably provided. At about 10 logic circuits per chip a new approach became necessary.

The next generation of integrated circuits were more complex but not so demanding in pins. To distinguish this class of circuit they are called *Medium Scale Integration (MSI)*. They have 10 to 100 logic elements per chip. We measure complexity in functional logic elements, rather than components, both to avoid excessively large figures and to give some feeling for what they will do.

Typical of MSI circuits are decoders and adders. These logic functions can be made from several logic gates interconnected together. They have a better ratio of complexity to pins, but can still be used as general-purpose building bricks by design engineers. Compared to the earlier levels of integration – now called *Small Scale Integration (SSI)* – systems made with MSI circuits have about ten times as much circuitry per unit area of printed circuit.

In the early 1970s the achievable complexity of a chip became sufficient to make functional units of more than 1000 elements. At this level the circuits are said to be *Large Scale Integration (LSI)*. This sort of complexity provided a breakthrough. All the important units of a simple electronic digital computer could be made as single integrated circuits.

Before the LSI level of complexity was reached a large number of SSI and MSI circuits were required to make the processing unit of a computer. The high-speed memories were core stores. They were mainly made from SSI and discrete-component circuits, together with a very large number of tiny rings of magnetic material; one per bit of memory.

LSI enabled the processing unit of a simple computer to be shrunk onto a single chip. A small memory could be made on another. This level of integration enabled the microcomputer to be created.

The early microcomputers were much less powerful than conventional computers, but continuous improvements in manufacturing technology have changed that. By 1980 integrated circuits containing over 100,000 logic circuits were being made. Above this level they are said to be *Very Large Scale Integration (VLSI)*.

Now, in the early 1980s, a small but complete computer can be made on a single chip. In sufficiently large quantities they cost less than an hour's wages! Nevertheless most microcomputer-based systems use components of all levels of complexity. Thus a typical printed circuit will be mainly populated by DILs. A few may have 40 pins, more will probably have 20 to 30, but a few will usually be SSI with 16 or fewer pins. When powerful output signals are required, there will also be some discrete components. Decoupling capacitors are always used. A

Q3.2–3.4, 3.6, 3.14–3.23   typical printed circuit is shown in Figure 3.10.

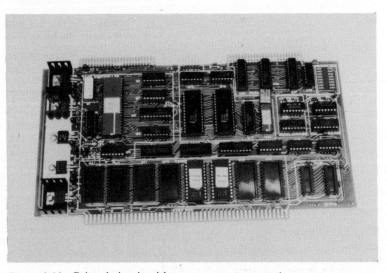

Figure 3.10   Printed circuit with components mounted

Questions

3.1   What are the most commonly used circuit elements?

3.2   What is an integrated circuit?

3.3   Are integrated circuits made from silicone?

3.4   What is a microelectronic circuit?

3.5   What are discrete components?

3.6   Do integrated circuits use resistors and transistors?

3.7   What does 10 kΩ mean?

3.8   What does 100 pF mean?

3.9   What would you expect a circuit component labelled R to be?

3.10   Sketch the symbols used to represent:
(a)   a resistor;
(b)   a capacitor;
(c)   a transistor.

3.11   Why are decoupling capacitors used?

3.12   Is a transistor a passive component?

3.13   Compared to bipolar transistors are MOS devices:
(a)   faster?
(b)   smaller?

3.14   Are printed circuits suitable for mass-production?

3.15   Why are printed circuits used?

3.16   The manufacture of printed circuits involves the use of a photographic film. Why?

3.17   What is photo-resist?

3.18   What is a silicon chip?

3.19   In what way is the manufacturing process for a silicon chip and a printed circuit similar?

3.20   What is a DIL?

3.21   What is a pinout?

3.22   Transistors have only three wires, DILS have more. Why?

3.23   What are the various levels of integration, and how are they distinguished?

Chapter 4 **Microcomputers**

Objectives of this chapter *When you have completed studying this chapter you should:*

1 Understand the general way in which computers work.
2 Be able to sketch a block diagram for a microcomputer.
3 Know that a program is a sequence of instructions.
4 Know that programs and data are stored in memories.
5 Be able to visualise information being transferred via a bus.

4.1 The stored-program computer

Microcomputers are electronic computers made from micro-electronic circuits in which the *processing unit is itself a single silicon chip*. Computers made from microelectronic circuits, but not based on a single chip processor, are not microcomputers.

All electronic digital computers work in the same way. Mainframe computers, minicomputers and microcomputers are all stored-program digital computers. For brevity we usually call them computers.

To avoid needless complexity we will examine the way in which computers work by reference to the very simplified schematic of Figure 4.1. It is based on a typical microcomputer and includes all of the essential features. The remainder of this section refers to it.

The computer contains four major units, connected together by a *bus*. This is an arrangement of wires that allows information to be passed between four basic units: the input unit, the memory, the micro-processor and the output unit.

The input unit

Low-level logic signals applied to the input unit are the systems inputs. They can be read, i.e. inspected, by the computer.

The output unit

Low-level logic signals are provided by the output unit. They can be set to either 1 or 0 by the computer. They are the system outputs.

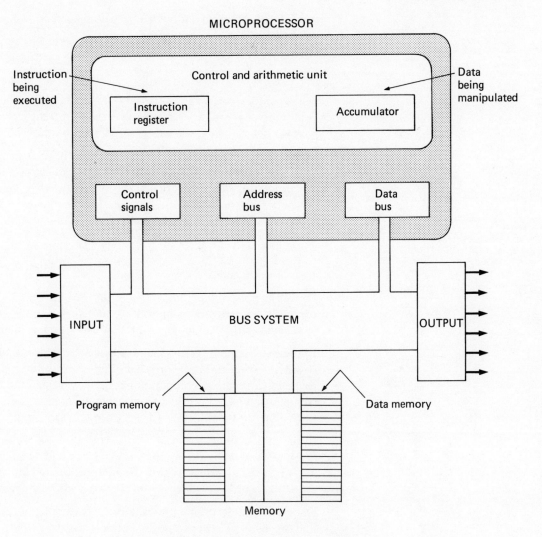

Figure 4.1 The stored-program computer

The memory

The memory is divided into two parts. One section stores the program and the other is used to hold information, or data required by the computer.

In a general-purpose computer the program is often changed. In a microprocessor-based system, such as a washing machine controller, the program is factory-installed and must never change. For this type of application the program is held in a special type of memory called a *ROM*. This stands for Read Only Memory. The computer is able to read the information but is not able to write to it, i.e. to change it. The program is then said to be *firmware*.

The microprocessor

A microprocessor is the processing unit for a microcomputer. All computers have one. It is the very heart of the computer. Its characteristics distinguish one computer from another.

Within a microprocessor may be over 10,000 logic elements. Figure 4.1 is very much simplified yet it shows the really important parts of any processing unit. In particular it contains a *control and arithmetic unit*. The control section tells the rest of the system what to do. The arithmetic unit, or more correctly the arithmetic and logic unit (*ALU*), performs any necessary computations.

How the units work

Having itemised the major units that together make a micro-computer, we will now examine how they work. First it is important to remember that computers represent everything as binary words.

In the program part of the memory is held a sequence of numbers (words) that represent the program. These words are called instructions. Each defines a computational task. Although the amount of computation performed by each instruction is very small, the micro-computer can obey a lot of them in a very short time.

In the data memory are held words that represent the information being worked on by the computer. The input and output signals are also arranged as words. Within the microprocessor chip are held the particular words on which the processor is actively computing.

Data is moved between units via the bus system; this will be described in more detail in a later section. At present it is sufficient to know that the microprocessor can gain access, i.e. get at, any stored or input data. It can also modify data stored in memory or change the outputs.

Computer action, as we described in Chapter 2, is a continual circular rhythm, of the form shown in Figure 4.2. It is usual to refer to these two phases as *fetch* and *execute*.

In the fetch phase the processor unit sends a message to the program memory asking for the next instruction. It does this by sending out control signals that mean *please supply an instruction* (from program memory). From the address controller it sends a number (the address) that indicates the storage location at which the next instruction will be found. Whatever instruction is stored at the addressed location in the program memory is read, i.e. copied, into the instruction register.

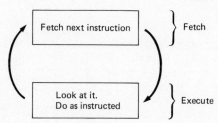

Figure 4.2 The basic two-phase rhythm of a computer

A *register* is a single word memory in which data is stored so as to be easily and quickly accessible.

Built into the microprocessor is an elaborate truth table. It includes all the possible instruction words. By comparing the binary code in the instruction register to that stored in the table, the computation to be performed is identified. Also built into the microprocessor are the necessary logic circuits to cause the instruction to be executed.

Sometimes an instruction can be completed within the microprocessing unit. Sometimes data must be obtained from input, or data memory. Sometimes data must be placed in memory or the outputs changed. In all cases the microprocessor causes the appropriate action to occur.

To help in the manipulation of data the processor includes some registers in which temporary results can be held. The most important of these is called the accumulator. In most computations there are two inputs and one output. Whenever possible one of the inputs is held in the accumulator and the output, i.e. the answer, is also written to the accumulator.

In this section we have examined the way in which all stored-program digital computers operate. Differences arise mainly in the variety of instructions that a particular computer can execute and in the time required to complete an instruction.

Q4.1, 4.2, 4.10, 4.14–4.19

4.2 The bus system

Information is passed between the various units of a computer via a communication subsystem called a bus. Over it a complete word can be transferred in parallel. Figure 4.3 shows the general arrangement as it is usually drawn. It is almost the same as Figure 4.1 except that the input and output arrangements are shown as a single unit. For brevity it is usually called the input/output, or I/O unit.

Notice that the schematic diagrams of microcomputers include bi-directional arrows. These two-way arrows indicate that information can flow in either direction. Sometimes data is fetched from memory to the microprocessor, sometimes the microprocessor sends

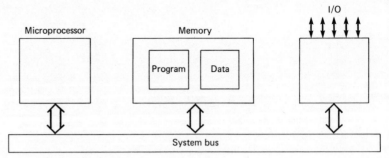

Figure 4.3   Schematic of a microcomputer

information to one of the other units. When data goes into the micro-processor it is said to be *reading*. When the microprocessor changes data stored in a connected unit it is said to be *writing*. Reference to Figures 4.4 and 4.5 will help reinforce this concept.

The bus system has two important components. They are:

- Interface circuits within each unit
- A set of parallel wires between all units.

Thus the various microelectronic circuits attached to the bus are all connected to the same set of wires. Figure 4.6 shows the general arrangement. Because traffic, i.e. data, passes along these lines the

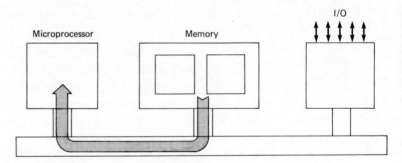

Figure 4.4   Reading from memory

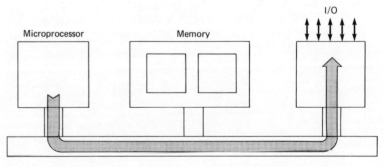

Figure 4.5   Writing to output

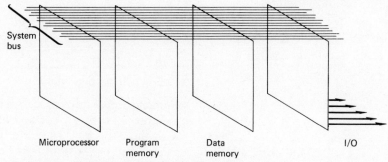

Figure 4.6 The BUS is a set of parallel wires

arrangement used to be referred to as a highway. Now we always call it a bus. In practice a microprocessor bus is divided into three groups of signals. They are:

- *The data bus* A set of wires, usually eight, along which data is transferred between units. Data can flow in either direction, i.e. it is bi-directional.

- *The address bus* A set of wires, usually sixteen, along which is sent the binary code (the address) that identifies the memory location to be used (see Figure 2.17). Addresses can only be issued by the microprocessor, so this bus is not bi-directional, it is uni-directional.

- *The control bus* A collection of special signals that coordinate the activities of the various units. Each type of microprocessor has its own particular set of signals, but in all cases they perform three important tasks. They are:

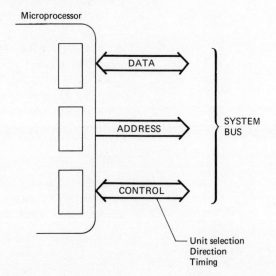

Figure 4.7 The system bus has about 38 wires: 8 data, 16 address and about 14 for control

1   *Selection*   These signals determine which signal is to be used.
2   *Direction*   This signal determines the direction in which data is transferred, i.e. read or write.
3   *Timing*   These signals define the precise times at which actions take place.

Summarising, a microprocessor bus is divided into three sets of wires, as shown in Figure 4.7. The control group of signals coordinate the activities of the unit. The address group selects a storage location and the selected data is transferred over the data bus.

Q4.11–4.13, 4.21–4.28

4.3   Memory

Within a computer information is stored in memories. The general principles of microelectronic memories were introduced in earlier sections but, because they are such an important part of a computer, we shall examine them in more depth.

In most computers the memory costs more than the microprocessor itself. This is because computers handle so much data. Many microprocessor-based controllers have less than 1 Kbyte of program. Some very clever computer programs are smaller than this. However, some computer programs occupy over 100 Kbytes. Moreover, in a general-purpose computer the programs are often changed, but they all have to be stored.

Even larger demands on memory space are made when there is a lot of information, i.e. data, to be stored and accessed. For instance, an airline booking system has to store all the time-tables, fare tables and all reservations for perhaps a few years ahead. So big computer systems have a hierarchy of stores. For long-term storage of large quantities of data we use magnetic media such as tape, or floppy disks, or Winchester disks. All these memories store the data as tiny magnetic patterns on a piece of material coated with a magnetic film. They are all based on the same principles as audio and video cassettes. To read or change the stored data the storage media has to be moving. In consequence it takes an appreciable time to get at a piece of data that is a long way away, e.g. at the other end of the tape.

Magnetic stores of the types mentioned enable us to store a lot of data cheaply. Unfortunately they involve moving mechanical parts and are very slow. In electronic terms it takes a long time to find things.

Microelectronic memories are made by the same manufacturing methods as microprocessors and other integrated circuits. They work at compatible speeds. Typically stored information can be retrieved or changed, i.e. read or written, in less than 1 microsecond (µs). This is at least 1000 times faster than can be achieved with continuously

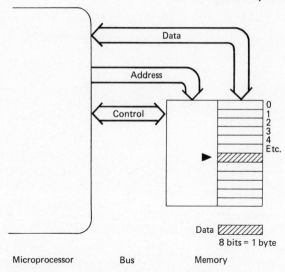

Data

Address

Control

0
1
2
3
4
Etc.

Data ////////
8 bits = 1 byte

Microprocessor   Bus   Memory

Figure 4.8   The address points to the selected location

rotating magnetic memories, such as floppy disks. It is over 1,000,000 times faster than a magnetic tape system.

Microelectronic stores are sometimes referred to as semiconductor memories. They are arranged as shown in Figure 4.8. It is more complete than Figure 2.17 and shows the three groups of bus signals.

To read from the memory the following sequence of actions takes place:

- Address is sent.
- Memory selects requested storage location.
- Control requests a read operation.
- Control causes data transfer via data bus.
- Control deactivates the memory.

Write operations are exactly the same except that data is transferred into the selected memory location. This causes a change in the stored data. Reading leaves the data unchanged.

A very important feature of the memories described is that the microprocessor can request stored data in any order. It can request data from location 98 and then from 83 and then from 108, etc. There are no restrictions on the ordering of successive addresses. Accesses can be in any order, it makes no difference. Such memories are said to provide random access. A *Random Access Memory* is referred to as a RAM.

By convention a RAM is defined as a memory to which we can write as well as read. In other words it can hold data that can be changed by the microprocessor. So a RAM is a random access memory that can store variable data.

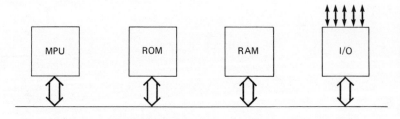

Figure 4.9   System schematic of most microcomputers

Because a RAM is an electronic circuit it needs power. If at any time, even for a very small fraction of a second, the power supply is outside specification, any stored data is lost. In particular, all the contents are lost when the system is switched off. For this reason the memory is said to be a *volatile store*. Switch-off causes the contents to 'evaporate'.

Microelectronic memories in which the stored data cannot be changed are called *Read Only Memories*, or *ROMs*. As with RAMs the data can be accessed in random order. By convention we call them ROM. They are mainly used to hold computer programs that must never change.·

As was hinted at the beginning of this section there are many types of memory. Besides ROM and RAM there are other microelectronic memories. An *EPROM*, for instance, is an *Eraseable Programmable ROM*. This allows a manufacturer to buy a standard chip and electrically program it. From then on it acts as a ROM but the equipment supplier has not had to purchase a minimum quantity of many thousands. Thus he can produce firmware, and therefore products, in small quantities.

Now that the different types of semiconductor memory have been introduced the schematic of a microcomputer can be redrawn in its usual form (see Figure 4.9). Permanant programs, i.e. firmware, are held in ROM. Variable data or temporary programs are held in **Q4.3–4.9, 4.20, 4.29**   RAM.

Questions

4.1 Is a microprocessor a stored-program digital computer?

4.2 Is a microcomputer a stored-program digital computer?

4.3 Name the four most important units of a microcomputer.

4.4 For each of the units given in answer to Question 4.3, briefly describe its purpose.

4.5 What form do the I/O signals take?

4.6 Does the read action change the data in the memory?

4.7 What do the following abbreviations stand for:
  (a) RAM?
  (b) ROM?
  (c) EPROM?

4.8 Which of the units listed for Question 4.7 are suitable for:
  (a) Holding firmware?
  (b) Random access?
  (c) Writing to?

4.9 Which of the memory units listed in Question 4.7 are non-volatile?

4.10 What is the purpose of an ALU?

4.11 Which part of a microcomputer controls the bus?

4.12 How is information represented within a computer?

4.13 To which parts of the system can the microprocessor gain access?

4.14 How does the processing unit find out what to do?

4.15 How many different instructions can be represented with one byte?

4.16 After fetching a new instruction, what must be done before the next one is fetched?

4.17 What are registers?

4.18 What is the accumulator?

4.19 What does a computer do:
  (a) Before the execution phase?
  (b) After the execution phase?

4.20 What is the purpose of an address?

4.21 Which part of the computer decides on the address of the next instruction?

4.22 Where are instructions held whilst they are being executed?

4.23 How many memory accesses are made during a single instruction?

4.24 How does the microprocessor know what to do?

4.25 What is the purpose of a register?

4.26 Do all stored-program computers work at the same speed and have the same repertoire of instructions?

4.27 Do you think that identical programs will run on different computers?

4.28 If a computer has an address bus with 16 wires, how big a memory can it have?

4.29 Draw a schematic of a microcomputer. Label it and briefly explain how it operates.

Chapter 5 **Peripherals**

Objectives of this chapter *When you have completed studying this chapter you should:*

1 *Understand what a peripheral is.*
2 *Know that a long-term memory unit is treated as a peripheral.*
3 *Understand that peripherals may not be physically separate units.*
4 *Be able to describe the purpose of the following popular peripherals: VDUs, teletypewriters, punched tape and floppy disks.*

5.1 Microcomputers and microprocessor-based systems

Peripherals are used to interface computers to the real world. They convert to and from the low-level electrical signals used by microelectronic circuits. In a very general sense they can be represented as shown in Figure 5.1. They enable a computer to input and output information. Examples of popular peripherals for computers are as follows.

Visual display unit

A visual display unit or *VDU* is used as an input device. Input is by a typewriter-like keyboard, output is by a TV-like screen. A typical VDU is shown in Figure 5.2. For office use they are the most popular peripheral. They are an essential part of a personal computer.

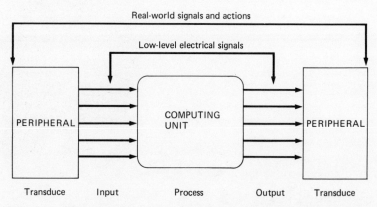

Figure 5.1 Peripherals are used to interface computers to the real world

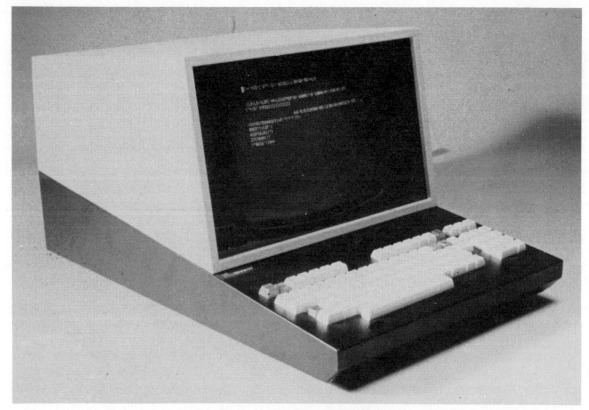

Figure 5.2   A visual display unit (VDU). *(Courtesy: Newbury Laboratories Ltd)*

Figure 5.3   A keyboard printer. *(Courtesy: Data Dynamics)*

Teletype

Teletypewriters*, like VDUs, are also input/output devices. Before microprocessors made VDUs cheap, teletypewriters were much used by computer programmers. Input is via a typewriter-like keyboard. Output, also like a typewriter, is printed paper. Figure 5.3 shows an example of the modern keyboard printer. The output can be controlled from the keyboard; alternatively a computer can take over.

Punched tape

Punched paper tape was used in all the early computers for inputing, outputing and long-term storage of data.

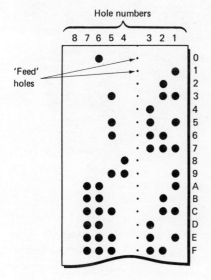

Figure 5.4   Punched paper tape

When new, paper tape is just a continuous narrow roll of paper. To store information a pattern of holes is punched across the tape. A hole represents a binary 1, no hole represents a binary 0. One row has up to eight holes and represents eight binary bits. This is enough for one character.

Notice that because the process of writing involves punching holes it is a destructive process. Tape can be written to only once. But it can be read many times. Perhaps because paper tape is so visible, storable and easily checkable, it is still used for some applications. Figure 5.4 shows what it looks like.

Teletypewriters usually include a punched tape facility. They can be used both to prepare and to read them into a computer.

Magnetic tape

Magnetic tape is also used to store binary data. Compared to paper tape it is much more effective. On one inch of magnetic tape can be stored almost one thousand times more information than on the same length of paper tape. Moreover it can store and provide data almost one thousand times faster. Whereas the peripherals previously described were obviously input and output devices, a magnetic tape unit is primarily a memory. But it is long-term memory. For this reason it is classed as a peripheral.

Unlike a semiconductor memory, storage media can be removed. Rolls of tape, or cassette units, can be manually filed and changed. Data stored in this way is also permanent. It is not lost when the computer is switched off, i.e. it is non-volatile storage.

Magnetic disks

Floppy disk units are now the most popular form of disk storage for microcomputers. They are based on a disk of magnetically coated

*Teletypewriter is a trade name of Teletype Corporation.

Figure 5.5   A selection of floppy disks

plastic. They are thin and flexible, hence fluppy. Figure 5.5 shows examples. Magnetic disk stores are based on the same physical principles as magnetic tapes. In comparison they are faster and more expensive.

In use the disk rotates continuously. In the time of one revolution any stored data can be read. It takes much longer to find data stored on a tape; it takes a long time to wind a long tape from one reel to another.

Transducers

All peripherals are transducers in that they convert from one physical form to another. However, as the preceding examples have shown, some computer peripherals are very complex mechanisms. Often they include several transducers. For instance, a teletypewriter includes transducers that convert:

- Keyboard motion to logic signals.
- Logic signals to paper movement.
- Logic signals to punch action.
- Detected holes to logic signals.
- Logic signals to print action.

So, whilst all peripherals are transducers, they are not measurement transducers. Nevertheless, computers are often required to make measurements; then measurement transducers are connected via converters (see Section 2.2).

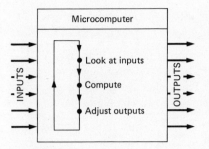

Figure 5.6 The microcomputer as a component

Q5.1–5.9

Summary

In summarising this section it will be helpful to first consider a microcomputer as a component. This viewpoint is shown in Figure 5.6. It shows the idea of a general-purpose component that continuously cycles through a three-part sequence of:

- Input
- Compute (process)
- Output

Because the computational part is defined by a stored program the component is very general-purpose. The snag is that the 1/0 is just a set of low-level electrical signals. The peripherals are the interface to the real world. They convert to and from low-level electrical signals to provide more convenient inputs and outputs. Often they dominate in both cost and size.

Most computer peripherals are physically separate units. They connect to the microcomputer unit by just a few wires. Because microcomputers are so very small the peripherals are usually much bigger than the computing electronics. Then the peripherals and the microcomputer are sometimes all built into one unit. Moreover, peripherals now often include one, or perhaps several, microprocessors. So it is only at the system schematic level that it is easy to identify a peripheral.

5.2 The VDU

A VDU is shown in Figure 5.2. They are suitable for desk-top use and provide a convenient method of communicating with a computer. Data is entered by using a typewriter-like keyboard. Each key stroke generates a binary word that tells the computer which key was hit. The ASCII coding system is used; it was mentioned in an earlier section.

Output is on a TV-like screen where information is displayed as lines of writing *(text)*. Usually a whole screen is divided into 24 lines, each of 80 characters. Thus on a screen we can display about half a page of normal typing. Thus a total of $24 \times 80 = 1920$ characters can be displayed at once.

Anyone unfamiliar with VDUs should endeavour to see, or better still, to use one. Preferably you should do this before studying the material that follows.

Communication between a VDU and a computer is by means of a serial communication link. This means that the information can be transferred along a pair of wires. This is convenient and cheap. To make it possible to interchange VDUs the serial arrangement is usually made to conform to an international standard (such as RS232).

To avoid the need for much attention from the computer, VDUs include a semiconductor memory. It is of a size sufficient to hold a complete screen-full of data. Thus the computer need only communicate with the VDU when there is a change of data.

When a key, 'z' say, is depressed, the VDU sends to the computer the binary code for 'z'. Nothing appears on the screen unless the computer sends a message to the VDU. So the 'z' is not automatically displayed.

When depression of a key causes the appropriate symbol (letter, number, etc.) to appear on the screen, the computer is said to echo the keyboard.

When a VDU is in use the computer to which it is connected is:

● Looking at the keyboard.
● Sending messages to the display.

In processing the keyboard data two tasks are being performed:

● Extracting input data for storage and further processing.
● Deciding what to display.

The computer program that looks after the VDU is often called a handler. It takes in data from the keyboard or from other programs and defines what is to be displayed.

A simple VDU handler would echo the keyboard and make the VDU very much like an electronic typewriter. Hitting 'abc' on the keyboard will cause 'abc' to be displayed. Hitting *carriage return* will cause a 'carriage return', i.e. a new line will be started. More impressively there will probably be one or more rub-out or delete keys. Hitting one of these will cause the disappearance of the last character to be entered, the last word, or even the whole line. This is a simple edit command. Such commands enable imperfect typists to create perfect text.

More sophisticated programs enable VDUs to be used in a more interactive way. In this type of operation the computer responds to the keyboard, quickly and intelligently. In this type of use the computer may be programed to ask the user what he wishes to do. In response it could cause the equivalent of a blank form to be displayed. The user is then expected to complete it, there and then, via the keyboard. A good program will check the data entered. If it is obviously silly the user will be made aware of this. A polite program might cause the VDU to display the code entered followed by a question mark. In come cases the program will refuse the data. This is easy to arrange, nothing happens; at least nothing appears to happen.

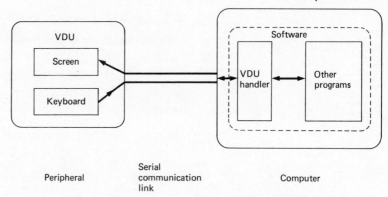

Peripheral   Serial communication link   Computer

Figure 5.7 VDUs act as a peripheral to a computer. The computer needs special software to handle it

In general a VDU is a general-purpose peripheral that enables people to communicate with a computer. VDUs are used in the way shown in Figure 5.7. Data is usually coded in the form defined by the American Standard for Information Interchange (ASCII). Besides text, visual displays can be used to display graphics and pictures. Figure 5.8 shows a screen displaying mixed text and graphic information for a security system. The display is actually in colour; the appropriate plan is automatically generated when an alarm occurs.

Q5.10–5.13

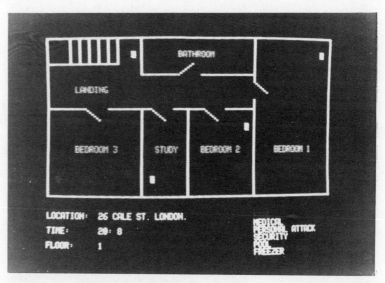

Figure 5.8 A graphic visual display for a security system. *(Courtesy: ARMA Systems Ltd and Cooke Associates)*

5.3   The teletypewriter

TTY is the accepted abbreviation for teletypewriter. Before VDUs became commonplace, TTYs were much used as input and output peripherals. The major differences are:

- TTYs work at mechanical speeds.
- TTYs produce hard copy.

Hard copy is the phrase used within the computer industry to refer to printed documents produced by a computer.

As the name implies, a TTY can produce typed paper. As with a VDU the keyboard produces binary codes to send to the computer. Received binary codes control the printing process. Only when the codes received are the same as those sent does the TTY act as a typewriter. Then it is said to be in local mode.

Traditionally a TTY was expected to include a paper tape facility. Then it could be used:

- To produce paper tapes from keyboard entries.
- To print out hard copy from paper tapes.
- To read paper tapes into a computer.
- To act as a computer-controlled printer.

Because teletypewriters are electromechanical, they are both slow and expensive. They are also unreliable. With the rapid disappearance of paper tape they are fast becoming a peripheral of

Q5.14–5.16   the past.

5.4   Punched tape

The idea of storing data on punched tape is fairly easy to explain because we can see it. It is easy to see the difference between plain tape and punched tape. It is easy to understand that each different pattern of holes can be used to represent a different symbol.

Punched tape is usually made of paper, but not always. Sometimes we use plastic tape, which is stronger and more durable, especially when used in oily surroundings. This occurs, for instance, when it is used for controlling machine tools in a factory.

To prepare punched tape, i.e. to write to it, we use an electronically controlled mechanism. It consists of mechanical punches that are electrically actuated by solenoids. These are devices that convert electrical energy into mechanical movement. The punch is driven by low-level, computer-compatible logic signals.

To assist in the accurate positioning of the tape the punch unit makes a series of small sprocket holes. They serve the same purpose as those on films.

A punch works according to the following algorithm:

- Advance tape to next position.
- Wait till new character received.
- Punch appropriate hole pattern.
- Repeat.

A paper tape reader works according to the following algorithm:

Do nothing until request to read achieved. Then:

- Read hole pattern (in parallel).
- Send the data serially (bit by bit).
- Advance tape to next position.
- Repeat.

In practice reading can be much faster than writing. Reading does not involve mechanical movement of a punch. At punching time the paper must be stationary. Reading is much easier. Often the holes are sensed optically. A light at one side of the tape is detected at the other by an electronic light sensor. This is a very fast process; the tape does not need to be stopped. It can be read whilst in continuous motion, i.e. on the fly.

From a systems point of view a paper tape unit can be visualised as performing the functions shown schematically in Figure 5.9.

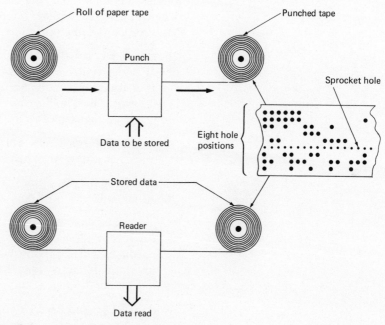

Figure 5.9 A paper tape unit

Important features of paper tape include:

- Durability
- Ease of storage
- Can be posted, etc.
- Computer-readable.

These features, combined with the fact that data is stored in a fairly obvious way, made paper tape very popular. For both long-term storage and moving data from one computer to another, paper tape was, for nearly 30 years, the preferred medium. Now it is too slow. Moreover the quantities of data now stored within computer systems would require impossibly large quantities of paper. Magnetic storage media, of one form or another, are now almost always used for long-term storage of computer programs and digital data.

Q5.17–5.19

5.5   Magnetic tape

Digital data can be stored on magnetic tapes. The principles are the same as those used to store music, etc., on cassettes. The storage medium, in both cases, is a long length of plastic tape coated with an extremely thin magnetic layer.

If the tape could be viewed under a special microscope it would show recorded data as a sequence of little magnets. Each would be less than a thousandth of an inch long. Each magnet represents one bit of data. When a 1 is stored the magnet is in the NS orientation. If a 0 it is in the SN direction, i.e. the other way round.

The stored data is really a magnetic field pattern. To read it the tape is pulled past a read head; this causes a pattern of induced voltage. Sensitive electronic circuits convert the tiny signals into a sequence of logic signals. Thus, pulling the tape over the read head generates an output signal that recreates the stored data.

Writing is performed in a similar way; the tape is pulled past an electronically driven write head.

The cheapest magnetic units use audio cassettes. They can store about one million bits of data on a single cassette.

Magnetic tape units designed especially for use with computers can store over ten million characters (80 million bits) on a single tape. As with audio cassettes, previously stored data can be overwritten. This is important. It means that the tape can be re-used and selected sections of data changed.

Whereas VDUs and the TTY are input/output peripherals, magnetic tape units are primarily long-term stores. They are used:

- To store large quantities of data.
- To retain data for long periods.
- To store data in computer-readable form.

When data is held in computer-readable form it is said to be *machine-readable*.

For brevity magnetic tape is often referred to as *mag-tape*.

Mag-tape is used to provide a backing store. We use this term to describe memories that back-up the high-speed semiconductor memories of the computer. Backing stores are peripherals that provide for long-term storage of large quantities of data in machine-readable form.

There is another important feature of mag-tape. Like cassettes, the *storage media* can be manually removed, stored and exchanged. In the computer room of a large mainframe computer installation one often sees several mag-tape units, together with several rack-holding reels of tape. In a sense these racks are the computer equivalent of filing cabinets.

5.6 Floppy disks

Floppy disks are another form of magnetic backing store. Instead of the storage medium being a long length of tape it is a disk of the same material. For ease of use it is protected by a stout plastic sleeve. Figure 5.5 shows what they look like.

Floppy disks, or floppies as they are often called, cost a few pounds each and can store over one million bits on just one side. Most floppies are *single-sided*, but some are *double-sided*.

To use a floppy disk it must be fed into a drive unit. Once inserted and the door closed, the disks are rotated, creating a spinning disk.

Data is stored on mag-tape as one or more *tracks* along the length of the tape. Each track has its own read and write head. With floppies the head is mechanically moved around. A microcomputer looks after this. Each track is a concentric circle. Floppies have the following important features:

- Each track is continuous.
- For each track there is just one head position.
- All data on a track passes the head once per revolution.

To find data stored on a tape can involve looking along the whole length of the tape. This takes quite a long time. With floppies, data can be found much more quickly. It involves:

- Positioning the head along the appropriate track.
- Waiting, at most, one disk revolution.

For these reasons floppies are very convenient to use.

At this point the perceptive reader may be saying 'but surely, one will not know which track to look at, so it would be necessary to look at all tracks'. In practice this very real problem has been solved. The computer is programmed to keep an index so that it knows which track to go to.

Floppy disk stores were developed and made cheap just at the time when microcomputers got to the stage of requiring a suitable backing store. Most microprocessor-based computers have two or more floppy disk units as peripherals. Figure 5.10 shows a typical microcomputer system. Similar equipment can be seen in any of the shops selling personal computers.

Figure 5.10   Microcomputer system including floppy-disk units. *(Courtesy: CASU Ltd)*

Initially floppies were all based on 8 inch disks. A large demand for even cheaper backing stores led to the development of a $5\frac{1}{4}$ inch unit. These are usually called mini-floppies. Most personal computers use them.

Q5.20–5.22   Mag-tape and floppies are just two of the types of peripheral based on magnetic storage.

5.7   Transducers

Transducers are a very important class of computer periheral. They are necessary in any system that measures or senses something.

A switch, giving an 'on' or 'off' signal is an example of an almost trivial peripheral. Despite its simplicity, such a device provides information that enables a computer to be programmed to respond to the real world; in this case to manual inputs.

Besides manually operated switches, transducers can be made to give a 1 or 0 signal depending on the sensed conditions. Examples include:

- The presence/absence of something.
- A door open/closed.
- A wire normal/cut.
- Satisfactory/contaminated.

Measurement transducers can also be connected to a computer. To do this we use a converter. These electronic circuits were introduced in Section 2.2. They convert analogue signals into computer-compatible digital data.

Measurement peripherals are not usually made as separate units. The transducer is connected to suitable electronic circuitry. Sometimes it is mounted on the same printed circuit unit as the microprocessor.

Transducers can also be used as output peripherals. For instance, computers can be programmed to produce electrical signals that look, and therefore sound, like audio ones. Fed into a loudspeaker, via suitable analogue circuits, computers can make music, or even synthesise speech. Electronic chimes and alarms use these ideas.

Q5.23, 5.24 In summary, computers can measure and control via transducers. Because they interface a computer to the real world (as compared to an electronic world), transducers are classed as peripherals.

5.8 Personal computers

When computers were large and expensive, a computer peripheral had its own big box. Now the actual peripherals to large mainframe computers may contain several microprocessors. At the other extreme some electronic equipment, or even toys such as dolls, can contain a microprocessor. It is therefore becoming more and more difficult to explain easily how to recognise a computer peripheral. Personal computers have made it even more difficult.

Personal computers are general-purpose computers based on micro-processors. They are desk-top size. To achieve this the necessary peripherals are often build into a single unit. The Hewlett Packard 85, shown in Figure 5.11, is a good example of this approach. Figure 5.12 shows its system schematic. Notice that the microcomputer has the following peripherals:

- Keyboard – for data entry.
- CRT – display screen.

Figure 5.11   The HP85 personal computer. *(Courtesy: Hewlett Packard Ltd)*

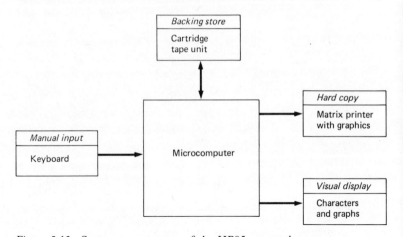

Figure 5.12   System components of the HP85 personal computer

- Tape cartridge – backing store.
- Printer – hard copy output.

Q5.25–28   Despite the fact that all the peripherals are in one box, the units connected to the microcomputer are still peripherals.

Questions

5.1   Why are peripherals used?

5.2   What is a VDU?

5.3 In what way is a VDU similar to a teletypewriter?

5.4 Why would you expect a teletypewriter to be noisy?

5.5 How is data stored on paper tape?

5.6 Is paper tape a non-volatile storage medium?

5.7 Is magnetic tape a non-volatile storage medium?

5.8 Can paper tape be re-used?

5.9 What storage media is used for:
(a) magnetic tapes?
(b) floppy disks?

5.10 How are inputs to, and outputs from, computers provided in a convenient form?

5.11 How do VDUs communicate with computers?

5.12 What is the output from a keyboard?

5.13 When a VDU is connected to a computer:
(a) How is the display controlled?
(b) How does the user enter data?
(c) What happens when a key is depressed?
(d) Will the screen display whatever is typed in?
(e) What is a handler?
(f) Why is ASCII relevant?

5.14 What is hard copy?

5.15 What are the outputs from a TTY?

5.16 List the possible inputs to a TTY.

5.17 Why does paper tape have sprocket holes?

5.18 Which would you expect to be the faster operation:
(a) Reading paper tape?
(b) Writing paper tape?

5.19 Can the data stored on paper tape be read without the use of a special reader?

5.20 In what form is data stored:
(a) On magnetic tape?
(b) On a floppy disk?

5.21 List three peripherals that can be used as backing stores.

5.22 The read/write head of a floppy disk drive is moved in response to messages from the computer. Why?

5.23 Why are measurement transducers that are connected to a computer called peripherals?

5.24   Why might a loudspeaker be interfaced to a computer?

5.25   Why do some peripherals include converters?

5.26   What peripherals would you expect to have in a personal computer system? Why?

5.27   In a microprocessor-based system, what would you use to:
(a)   Input information?
(b)   Output information?

5.28   What are the most popular peripherals used for computer games?

Chapter 6 Microcomputer hardware

Objectives of this chapter *When you have completed studying this chapter you should:*

1 *Be able to sketch a programmer's view of a microcomputer, showing microprocessor, bus, ROM, RAM, I/O, registers.*
2 *Understand the purpose of the program counter.*
3 *Know that the instructions available to the programmer allow movement of data, manipulation of data and changing the sequence of instructions.*
4 *Be able to understand some real instructions.*
5 *Be able to start using manufacturers' data.*

6.1 The programmer's view of microcomputer hardware

The way in which the stored-program computer works was introduced in Chapter 4. In this section we will review and extend these ideas and examine the way in which simple programs operate. Finally you will be asked to prove you acquired knowledge by writing a program and making it work. To make this possible the simplified schematic of Figure 6.1 (a reproduction of Figure 4.1) must have a few features added to it. It must also be redrawn so that it is easier to visualise what is happening as the program proceeds.

A system schematic sufficient for the remainder of this book is shown in Figure 6.2. It is next to Figure 6.1 so that you can easily compare the two schematics. *You should study them and note the similarities and differences*. In particular notice:

- *ROM and RAM* Together RAM and ROM are the memory, or store, of the computer. Each of the small rectangles is meant to convey the idea of separate storage location. By showing them like a ladder, the idea is to indicate that they are a succession of adjacent addresses (1, 2, 3, 4, etc.). The arrow is meant to indicate an address pointer. The address comes from the system bus, which includes the address bus.
- *I/O* This is addressed in much the same way as memory. The diagram reflects this. It also indicates that there are wires connected to the I/O. The memory does not have I/O wires. Furthermore there will be a very large number of storage locations, but few I/O *ports*. One byte of I/O is called a port. To program it appears as a byte of storage.

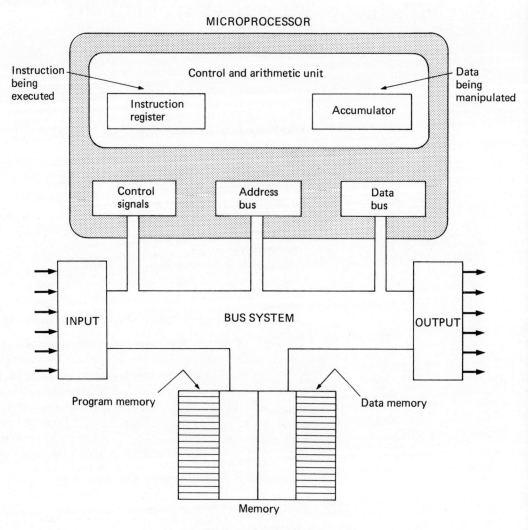

Figure 6.1 The stored-program computer

- *More registers* Compared to Figure 6.1 the microprocessor has more registers. Real microprocessors have many more, but Figure 6.2 includes as many as we need to follow through simple programs.
- *Different size of registers* In an 8-bit microprocessor the data is handled in 8-bit units. Nevertheless the address will be 2 bytes, i.e. 16 bits wide. The diagram reflects this.
- *Register* When writing programs we try and keep data that is being worked on within the microprocessing unit. Data held in this way is easily and quickly available. The registers provide this facility. Real microprocessors have many registers. At this stage

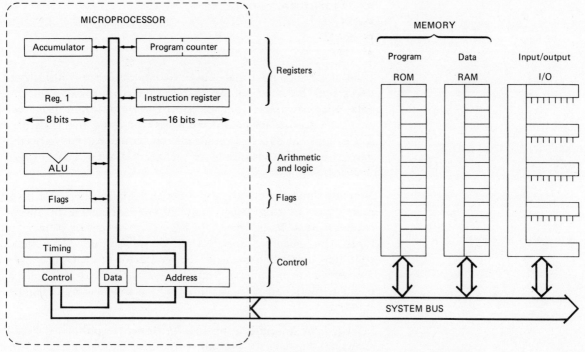

Figure 6.2 A programmer's view of a microcomputer

it is sufficient to have just one data register in addition to the accumulator.

- *Flags* The idea of using a single bit of memory as a flag was introduced in an earlier section. Real microprocessors have several flags. They are collected together in groups of 8. The diagram shows one set.
- *Program counter and incrementer* The *program counter* is a very special register. It holds the instruction address. It is called a *counter* because it is normally *incremented by one* during each instruction. That job is done by the *incrementer*. This is not an ordinary register; it is also a counter.
- *System bus* As was explained in Section 4.2 the system bus is divided into three parts: address, data and control. This is reflected in the diagram.
- *Internal bus* To move information around within the microprocessor is an *internal bus*. Real microprocessors sometimes have several of them. The diagram indicates the general idea.

Q6.1, 6.2

6.2 Microcomputer components

Microcomputers are made from microelectronic components of which the most important are:

- The microprocessor
- ROM
- RAM
- I/O

Each of these components is a very complex device, but sufficient background has now been covered to start referring to *manufacturers'* data. Selected extracts from published data are given in Appendices 1 and 2. In all cases the data is *much* less than is required to define a single chip! The data in a simple microcomputer extends to over 70 pages. For a more advanced type the specification of a *single chip* is thicker than this book!

Note also that there are many different manufacturers and many more types of chip. Some popular types are made by several manufacturers. Moreover, there is a very wide choice of memory types. The appendices merely show a cross-section. The author's collection of data books extends to over 2 metres of shelf space, and is nowhere near complete! You are *encouraged* to get hold of manufacturers' data with the aim of getting used to using and understanding it.

Refer first to Appendix 2, pp 161–70. Notice the pin-out. It clearly shows the bus signals. Not all of the descriptive material on the control signals will be meaningful. However, you should at least find that your vocabulary now includes some of the specialist words and abbreviations. Notice particularly the entries for address bus, data bus and reset. Do not be put off by $\overline{reset}$. The upper line means NOT. Thus $\overline{reset}$ = not reset, i.e. the binary opposite of reset. When reset is a 1, $\overline{reset}$ is a 0, and *vice versa*.

Refer next to Appendix 1, page 142. Shown is the schematic for a complete microcomputer. It includes:

- 8085, the microprocessor
- Memory
- I/O

Comparison with Figure 6.2 should show a strong resemblance. The device shown connected to pins X1 and X2 of the 8085 is a quartz crystal. It is the same type as is used in a quartz watch. It defines precisely the very high-speed clock frequency for use by the whole microcomputer system. A typical frequency is 4 MHz (4 million oscillations/second).

The schematic for the Z80 is shown on page 161 of Appendix 2. Compare it with Figure 6.2. Compare also the more detailed diagram of the 8085 shown on page 142 of Appendix 1.

We shall now return to considering the way in which all microcomputers work. Once again Figure 6.2 will be our model. It is the simplest we can get away with. By making reference to data on actual devices you will probably have realised that:

Q6.3–6.6
- Simplified models are vital to understanding the ideas.
- Real devices are very complex.

6.3 The program counter

All stored-program digital computers have a *program counter*. It is a very special *register*. It controls the sequence of instructions followed by the computer. By the end of every instruction the program counter contains the address of the next instruction.

At the beginning of each new instruction the contents of the program counter are used as an address. From this address in memory the next instruction is fetched. Once the new instruction is within the processor it is stored in the instruction register. Finally the contents of the program counter are changed. It is made equal to the address of the next instruction. Stated more briefly; the program counter is made equal to the address of the next instruction. An even more concise way of saying the same thing is:

(Program counter) ← Address of next instruction

In this example a standard convention is used. Brackets surrounding the register name mean *the contents of*. Thus (REG 1) means the contents of REG 1. This method of expressing computer actions is widely used. It is important that you become familiar with it.

In the simplest programs the instructions come from successive addresses. Then the program counter proceeds in a sequence, such as 7, 8, 9, etc. That is why the register holding the program address is called a program counter.

When successive instructions come from successive addresses the action that takes place for each instruction can be expressed as shown in Figure 6.3. In this example PC is used as an abbreviation for program counter. Usually computers are arranged to fetch the first instruction from address 0. The next will be at address 1, the next at 2, and so on. Thus the program counter is usually *incremented* by one.

In the last paragraph a very important feature was glossed over. When a computer is first switched on it is vital that we know what it will do. So we must know the address of the first instruction that it will fetch. For this reason all computers have *reset* facility. When first switched on, or whenever the reset button is pressed, the computer is forced to fetch its next instruction from a particular address. Usually it is address zero. This action can briefly be stated as:

Reset causes   (program counter) ← zero

(PC) ← (PC) + 1

Contents of program counter

Contents of program counter + 1

Figure 6.3 Abbreviated way of stating that the contents of the program counter (PC) are incremented

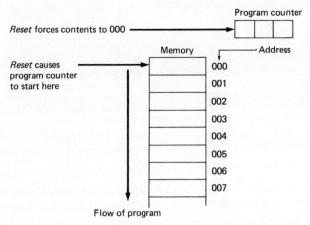

Figure 6.4   The simplest programs are sequential; reset starts execution at address zero

Because we must often refer to what happens when the equipment is first switched on, we talk of automatic reset being *reset on power-up*. When a program proceeds one instruction after another, we can represent the situation in the way shown in Figure 6.4. Reset causes the program counter to be forced to zero, i.e. the contents are made zero. When reset is released the control unit uses the contents of the program counter as the address of the next instruction. After the first instruction has been completed the control unit, once again, uses the contents of the program counter as the address of the next instruction. The whole program repeats itself over and over again in the way summarised in Figure 6.5.

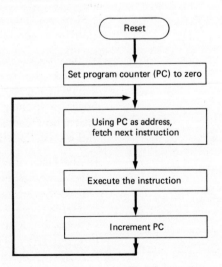

Figure 6.5   Flowchart for a simple (sequential) program

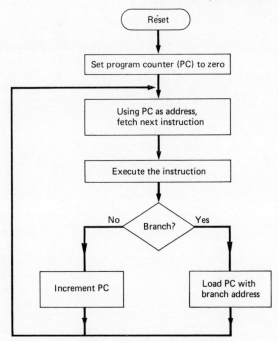

Figure 6.6 Basic algorithm followed by all computers

In practice, computer programs are more complex than just a straightforward sequence such as 8, 9, 10, 11, etc. They include loops for instance. To enable this there is a special class of computer instruction. Their purpose is to change the flow of instructions; we call them *flow-of-control*, or *branch* instructions. The simplest example of this class of instruction is the GOTO type. If, for instance, the instruction at address 35 uses GOTO 34, the sequence of instructions will be 34, 35, 34, 35, 34, 35, etc.

All computers have branch instructions. Figure 6.6 shows the flow-of-control arrangements obeyed by *all* stored-program digital computers. When a branch instruction is encountered the program counter is not incremented; it is loaded with a new address. Often this new address is part of the branch instruction, i.e. GOTO 34.

Instructions that cause the program counter to jump out of its normal sequence are usually called JUMP instructions. For brevity JUMP to 52 is usually written as JMP 52.

In a typical computer there will be several types of instruction that can effect the flow-of-control. A very important class are referred to as *conditionals*. These instructions may, or may not, direct the program counter from its normal sequence; it depends on a flag. An example of this type of instruction is:

```
JMP to XX if F2 = 1
```

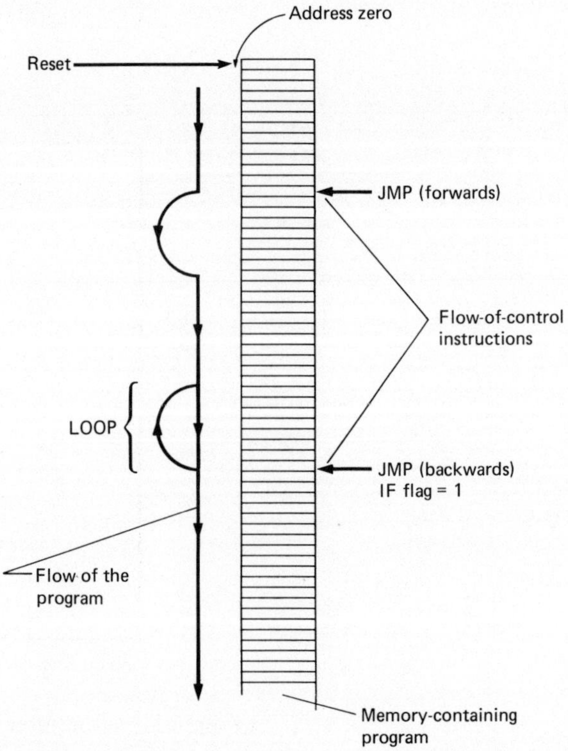

Figure 6.7 Flow-of-control instructions allow the programmer to have alternative actions, loops and repetition

If, when this instruction is encountered, the flag F2 is a 1, then the next instruction will be that at address XX. If, on the other hand, the flag F2 is not a 1, i.e. it is a 0, then the program counter will be incremented. The next instruction will then come from the adjacent address.

Using these flow-of-control instructions the program can be made to *jump around* in a controlled manner. Instead of the simple progression shown in Figure 6.4 the more complicated pattern of Figure 6.7 can be arranged. It includes an unconditional jump. Then the program counter is made to jump over some instructions, regardless. A backwards jump creates a loop.

Unless a conditional instruction is enclosed within a loop, they never end! By including a test, the program is made to stop looping when some particular condition occurs.

In this section we have studied the way in which the program counter controls the way in which a program is executed. To consolidate these ideas we will now follow through a simple program. To do this we will

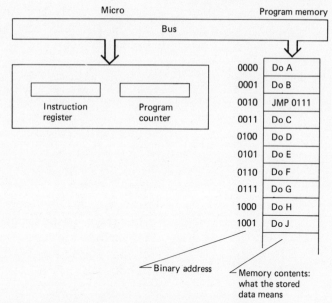

Figure 6.8 Schematic of a program stored in memory

use the simplified model of Figure 6.8 showing:

- The program counter PC.
- The instruction register IR.
- Part of the program memory.

```
Following reset we have

        IR            PC
       ────          ────
       XXXX          0000

Note XXXX is used to indicate 'don't care'.
The (PC) is used as an address.
It points to the instruction code (for Do A).
It is loaded into IR
Producing

        IR            PC
       ────          ────
       Do A          0000

The instruction is now executed (obeyed).
Then the (PC) is updated (incremented).
Producing

        IR            PC
       ────          ────
       Do A          0001

The (PC) is used as an address.
```

It points to the instruction code (for Do B).
It is loaded into IR
Producing

IR	PC
Do B	0001

The instruction is now executed.
Then the (PC) is updated.
Producing

IR	PC
Do B	0002

The (PC) is used as an address.
It points to the instruction code for JMP 0111.
It is loaded into IR
Producing

IR	PC
JMP 0111	0002

The instruction is now executed.
It causes no computation but.
The (PC) is updated to 0111
Producing

IR	PC
JMP 0111	0111

The (PC) is used as an address.
It points to the instruction code for Do G.
It is loaded into IR
Producing ..... and so on.

In this example the computer has been made to follow a program that starts by doing tasks A, B and then G, etc. If the instruction at address 0010 were changed to JMP 0101 the parts performed would be A and

Q6.7–6.13   B, and then E, F, G, etc.

6.4   The instruction set

Built into a microprocessor is a control unit that can obey a range of tasks defined in the *instruction set*. Typically there will be a repertoire of about 100 different instructions. Each instruction is a binary word of one or more bytes.

At this stage it is convenient to consider instructions as being of three types.

- *Data movement*   Instructions of the data movement variety are used to store, retrieve and move around data. An example of this type is:

```
Load the accumulator from address 99
```

- *Data manipulation*   Instructions of the data manipulation type cause a *change* in the stored data. Examples are:

```
Add A to B
If A = B set flag to a 1
```

- *Flow-of-control*   Instructions that can divert the program counter from its sequence are called flow-of-control instructions. They were introduced in the last section. Examples are:

```
JMP to 83. If flag = 1
JMP to 45. If flag = 1
```

The first two classes of instruction always involve two addresses and often three. For instance the instruction 'Add A to B' requires three addresses:

1   One address to define where A is.
2   One address to define where B is.
3   One address to define where the answer is to go.

In other words there are two *sources* and one *destination*.

If each address requires 16 bits, i.e. two bytes, that would mean that attached to each add instruction would be three addresses each of 16 bits! Because this is excessive we use registers in a variety of ways. We also define the address in many different ways. In this course we can only begin to introduce the simplest and most basic of these techniques. The topic is generally discussed under the heading of *addressing modes*.

Appendix 1 reproduces details of the instruction set of the Intel 8085 microprocessor. By studying the *summary of processor instructions* you will notice that each instruction is given a mnemonic. Each instruction is represented by just a few letters. They are carefully chosen so that their meaning is as obvious as possible. Notice for instance that all the instructions that start with J are some variety of Jump. Notice that

JZ means Jump on Zero
JNZ means Jump on Not Zero

Similarly:

CALL means Call Unconditionally

CC means Call on Carry

CNC means Call on No Carry

At this stage it does not matter what call means.

In studying these instruction codes for the 8085 notice that the binary code is given. For instance:

11111010 is shown as meaning JM (Jump on Minus)

In other words the instruction code for JM is 11111010. From this example it will already be clear that the mnemonic JM is given for us. Binary code is best handled by computers and avoided by people!

Unfortunately computers differ in their instruction sets. Even worse, instructions that do the same thing may not have identical names. This creates a problem. Only by writing programs for a real computer can they be tested, but not everyone studying this book will have access to the same type of computer. It is for this reason that manufacturers' data has been included. By following the descriptions of a small selection of real instructions for the 8085 you should be able then to understand similar instructions from the published data of other microprocessors. To prepare a few simple programs we need to use only a small selection of the instructions that are available. We will now study a set that covers the following operations:

- Load
- Store
- Add
- Increment
- Compare
- Halt
- Jump – unconditional
- Jump – conditional

By reference to the data on the Intel 8085 reproduced in Appendix 1, we will now study some real instructions that cover the above range.

Data movement

When data stored in one place is wanted in another we talk of *moving* it. In fact we

- Read from the source.
- Write to the destination.

The source is left unchanged.

As Figure 6.9 shows there are two basic types of operation. When data is moved from memory to processor it is said to be a *load* operation. When data temporarily held within the processor is moved to memory, it is said to be *stored*.

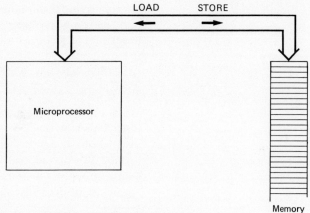

Figure 6.9 The LOAD and STORE operations

A third class of data movement instruction is usually provided in that data can be moved between registers within the microprocessor. The 8085 has the instruction MOV which is mnemonic for *move register*.

Within the processor are several registers. The accumulator is called the A register. There are also B and C registers. Reference should now be made to the appropriate parts of Appendix 1. Notice that the instruction code starts with 01 and is followed by DDD and then SSS. These letters indicate that they must be *replaced by the appropriate pattern*, of ones and zeros. Thus the transfer of data from register B to register A requires the binary code 01111000. By studying Appendix 1 and Figure 6.10 you should understand why this is so. You should test your understanding by making up the instruction code for some other register-to-register moves, e.g. A to C.

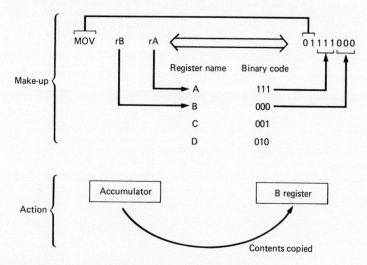

Figure 6.10 The MOVE A to B instruction (MOVr1,r2)

Before leaving this example notice that it involves two addresses. In both cases registers were involved. There are only a few of these so they could be identified with first three bits each. The first two bits of the instruction – in this case 01 – defined what type of operation had to be done. The next six bits identified, i.e. addressed, the source and destination.

When the memory is involved the address required is usually 16 bits long. This is twice as many bits as was needed for the entire MOV instruction! In consequence instructions that involve the memory include only one address. For a *load* type of action the memory is the source. A register within the microprocessor is the destination. In the *store* type of operation it is the other way round.

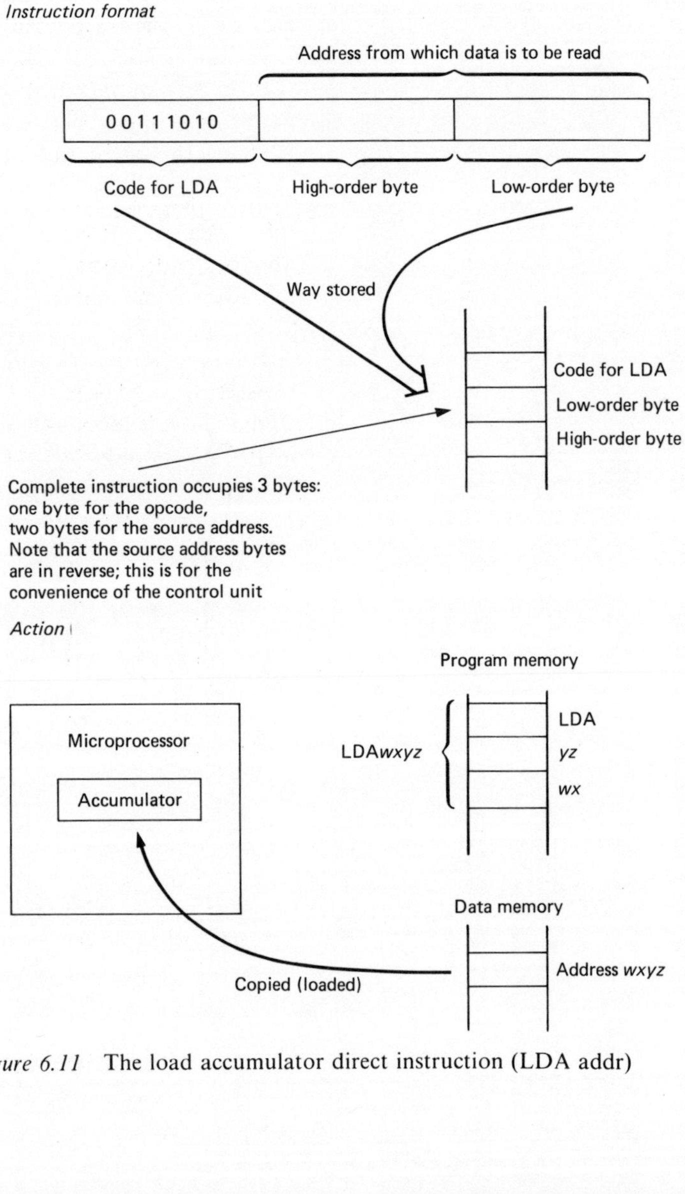

Instruction format

Address from which data is to be read

0 0 1 1 1 0 1 0		

Code for LDA   High-order byte   Low-order byte

Way stored

Code for LDA
Low-order byte
High-order byte

Complete instruction occupies 3 bytes:
one byte for the opcode,
two bytes for the source address.
Note that the source address bytes
are in reverse; this is for the
convenience of the control unit

Action

Program memory

Microprocessor

Accumulator

LDA*wxyz*

LDA
yz
wx

Data memory

Copied (loaded)

Address *wxyz*

Figure 6.11 The load accumulator direct instruction (LDA addr)

The 8085 instruction LDA is typical of the type of operation. It is what is called a *Load accumulator direct* action. Reference to the appropriate parts of Appendix 1 and to Figure 6.11 will help make its meaning clear.

The example introduces a number of unavoidable complications:

1 Firstly the instruction is *three* bytes long. The MOV instruction was only *one* byte long. This means that the program counter must be incremented not just by one, but according to the length of the instruction. Having noticed the problem, we can now forget about it. The control unit within the microprocessor automatically takes care of this. It knows what instruction is being executed, it knows how long it is and acts accordingly.
2 There is no choice of destination. The data is automatically loaded into the accumulator.
3 The address is stored as part of the program. In the case of the 8085 the two bytes that together form the address are stored lowest-order byte first. This is not the way round we would naturally tend to store them. Not all microcomputers put them this way round. But in the 8085 that's the way it is.

To make the last point clear, let us consider the instruction that requires that the accumulator be loaded from binary address 1111111100000000. In memory the instruction would be stored as:

```
Byte 1    00111010
Byte 2    00000000
Byte 3    11111111
```

Notice that the total instruction has three bytes. The first byte defines the operation to be performed. It is called the operation *code*. For brevity this is almost always abbreviated to *opcode*.

When the opcode is followed by an address or data it is said to be an *extended* instruction. Thus the LDA instruction has an opcode of one byte followed by an address part of two bytes.

Before leaving the LDA instruction it will be useful to follow through the sequence of actions that occur. They are:

- *Step 1*   Instruction is fetched (loaded) and inspected.
  Instruction code for LDA found.
- *Step 2*   Next two bytes loaded (into processor).
  Placed side by side to form 16-bit address.
- *Step 3*   Address used to access memory.
  Data found is loaded into accumulator.

A final comment on the example concerns it being called load accumulator *direct*. It is called direct because the instruction includes the address. In more complex addressing modes the instruction only says *where* the address is to be found.

Our next, and final, example of the data movement type is the instruction 'store accumulator direct' which is abbreviated to STA. It is a three-byte instruction very similar in make-up to the one we have just studied. The action is defined in Appendix 1 and is repeated below:

$$((\text{byte } 3)\ (\text{byte } 2))\ \leftarrow\ (\text{A})$$

At first sight this will probably seem off-putting. Reference to Figure 6.12 will help to make the attraction understandable.

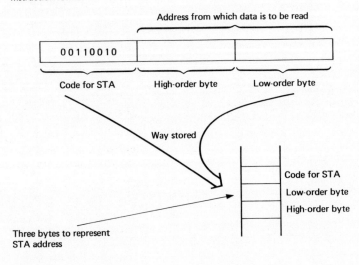

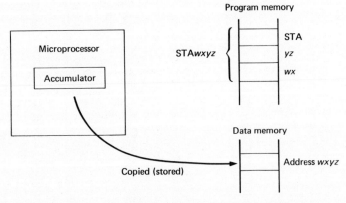

Figure 6.12   The store accumulator direct instruction (STA addr)

In Appendix 1 the manufacturer describes the instruction as follows: 'The contents of the accumulator is moved to the memory location whose address is specified in byte 2 and byte 3 of the instruction'. If

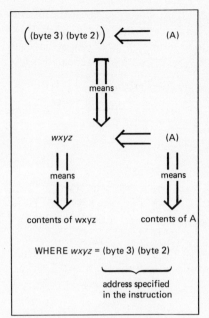

Figure 6.13 Definition of the STA instruction

the bracketed expression is read in the way explained in Figure 6.13 it will become clearer. Only practice will make it clear.

Data manipulation

Arithmetic instructions are typical of the data manipulation class. They act on one or more numbers to produce an answer. Usually the answer is held within the microprocessor.

Often a computer program will do the following:

- Load data
- Manipulate it
- Store the answers

In the 8085 instruction set we will first study ADD r. This stands for *ADD register*. The action performed is:

$$(A) \leftarrow (A) + (r)$$

Figure 6.14 also explains the action performed. The instruction format is similar to MOV, except that only one register must be specified. The answer *and one* of the inputs always comes from the accumulator.

By studying Appendix 1 you should be able to deduce the correct instruction code for adding the contents of any register to the accumulator.

Action

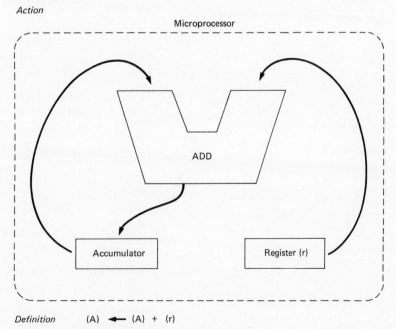

Definition $(A) \leftarrow (A) + (r)$

Figure 6.14 The ADD register instruction (ADD r)

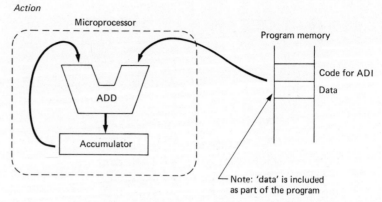

Figure 6.15   The ADD immediate instruction (ADI data)

Another similar instruction in the 8085 set is *add immediate* which has the mnemonic *ADI data*. It is a two-byte instruction. It is defined as:

$$(A) \leftarrow (A) + (byte\ 2)$$

Figure 6.15 shows what happens. Whenever data is *attached* to the instruction it is said to be *immediate*.

A rather different type of instruction is the increment/decrement family. This sort of instruction has one input and one output. The best example in the 8085 set is *decrement register* or DCR r. It performs the following action:

$$(r) \leftarrow (r) - 1$$

If, for instance, the specified register contained the binary equivalent of seven, then the decrement instruction would cause the contents to change to six. The action of DCR is shown in Figure 6.16.

A rather different sort of data manipulation instruction is *compare*. The 8085 instruction set includes CMP r. It leaves *unchanged* the contents of both the accumulator and the specified register.

Instructions of this type are used to set flags. Then conditional branch instructions can be used to inspect the flags.

The instruction CMP sets several flags. At present it is sufficient to know that:

If $(A) = (r)$, then Z flag is set to 1
If $(A) < (r)$, then CY flag is set to 1

Thus the possible outcomes of a CMP instruction are

	Z	CY
Accumulator contains bigger number	0	0
Numbers equal	1	0
Register contains bigger number	0	1

Action

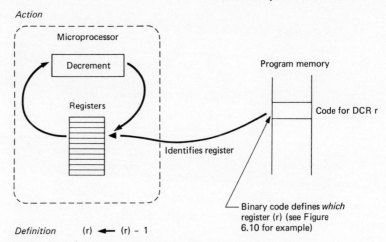

Definition   (r) ← (r) – 1

Figure 6.16   The decrement register (r) instruction (DCR r)

Flow-of-control

Flow-of-control instructions were introduced in the last section. Now we will look at some real examples. First we will examine the 8085 instruction JMP addr. This is an unconditional jump. It is defined as:

(PC)   ← (byte 3) (byte 2)

Once again it is a three-byte instruction. The first byte of the instruction is the binary code meaning JMP. The next two bytes are the address to which the program is to jump. The action is summarised in Figure 6.17.

The conditional jump instruction for the 8085 can inspect any one of eight flags, including Z and CY. The instruction is fully explained in Appendix 1 and summarised in Figure 6.18.

Action

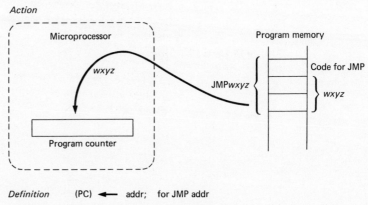

Definition   (PC) ← addr;   for JMP addr

Figure 6.17   The jump instruction (JMP addr)

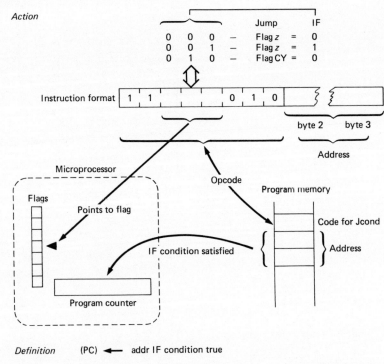

Figure 6.18   The conditional jump instruction (Jcondition addr)

Finally in our study of real instructions we will examine an instruction of the control type. The 8085 includes the instruction HLT. It is a single-byte instruction and is defined in Appendix 1. It halts the processor.

Halt instructions are sometimes a convenient way of making a microcomputer wait, without computing. For instance, even a very slow microprocessor will be able to execute over 100,000 instructions per second. Simple programs, such as you will be asked to write in the next chapter, will take only a tiny fraction of a second to run. The halt instruction will enable your program to stop the computer when it is done.

Q6.14–6.17

(*NOTE:* Question 6.16 is *very important*. You should not proceed to Chapter 7 until you have completed Question 6.16 and then properly done 6.17.)

Questions

6.1 Refer to Figure 6.2. Make a list of all the important parts, and *briefly* describe their purpose.

6.2 What is it that determines the address from which the next instruction is fetched?

6.3 What does enable mean?

6.4 Why do microcomputers use quartz crystals?

6.5 What does (*accumulator*) mean?

6.6 What is the purpose of Re-set?

6.7 Produce a flow chart that describes the way in which the sequence of instructions is controlled.

6.8 What is a branch instruction?

6.9 Describe the purpose of the jump instruction.

6.10 Explain what would happen if, at address 63, there was the instruction JMP 62?

6.11 Discuss the problem produced by the program suggested in the previous question.

6.12 What is an *instruction*?

6.13 What is an *instruction set*?

6.14 The instruction set can be conveniently divided into three types of operation. Name them.

6.15 Using the answer to the previous question, classify the following operations:
(a) Load
(b) Store
(c) Add
(d) Decrement
(e) Compare
(f) Halt
(g) Jump

6.16 Complete the instruction codes for each of the instructions shown in the blank table given overleaf in Figure 6.19. Ideally you should take a copy and use it as a blank form.

6.17 Some of the instructions used in the preceding question were described in this chapter, but some are new. For any of the instructions that you are a little unsure of, produce a sketch to show what the instruction does.

Instruction	Op-Code									byte 2	byte 1
MOV A B											
MOV B A											
MVI A data											
MVI B data											
LDA addr											
STA addr											
ADD B											
ADI data											
INR A											
INR B											
DCR A											
DCR B											
CMP B											
CPI data											
STC											
JMP addr											
JC											
JNC											
HLT											
NOP											

Figure 6.19

Chapter 7 Programming

Objectives of this chapter *To complete this chapter you must use a computer. You must write, run and observe some simple programs written in machine language. To do this you must have a suitable computer. On completion you should be able to:*

1 *Enter a program.*
2 *Run a program.*
3 *Use a simple assembler.*
4 *Use manufacturers' data to help understand new instructions.*
5 *Write a simple program.*

7.1 Programming facilities

This chapter, besides being the last of this book, is rather different to the earlier ones. It is very much concerned with helping you to write and run a few simple programs. To do this you will need access to one of the many types of suitable microcomputer. The essential features required are:

- Keyboard
- Display
- Assembler or machine code facilities
- Operating system or monitor
- Documentation
- Preferably a printer

In addition you will find it *most helpful* if you can gain the advice of someone who is already familiar with the system.

Notice the need for documentation. It is *essential* to have a reference manual that defines what the equipment can do and how to make it do it.

All the other requirements should be meaningful to you, except the need for an assembler or machine code facility, and also for an operating system or monitor. Both these items are special pieces of software. The most suitable computers for your programming work will have them built in.

A computer with BASIC is not in itself satisfactory. Some computers have BASIC and assembler but just the former is not enough. The difference is as follows. BASIC is what we call a high-level language.

The instructions we use in BASIC are much more powerful than the actual computer instructions that we wish to deal with. In fact a high-level language is a special piece of software. For each high-level language statement, a whole sequence of machine-language instructions are executed.

By contrast an assembler is a special piece of software that helps us to write programs in machine language. It lets us, for instance, write STA followed by an address when we wish to use the STA instruction. It avoids us having to discover that STA is really the binary byte 00110010. The operating system, or monitor, as it is often called in very small computers, is another special piece of software. Its job is to let us control the computer from a keyboard. For instance, it will let us enter our program into the memory, and let us inspect the results of running our programs.

Unfortunately there is very little standardisation in the computer world. Firstly there are a number of fundamentally different micro-processors from which the microcomputer may be built. In practice the most likely possibilities are the:

- Intel 8085
- Zilog Z80
- MOS Technology 6500

Additionally the manufacturer of the system will impose his own particular personality on the computer system. For instance, the PET, the APPLE, the AIM 65 and the Acorn all use the 6500. To the average user this is not at all obvious.

The reason for highlighting these differences is that it is only possible to write very precise instructions on how to run a program for just one particular computer. The alternative – to try and describe what to do for any and all computers – is not practical. In consequence, in writing this unit, a problem had to be solved. The chosen solution is as follows.

In the last chapter the Intel 8085 instruction set was used as an example. Anyone using a Z80 system will find the instructions described are all included. In fact the Z80 instruction set includes all of the 8085 set; there are also some additional ones. All you must watch out for is the use of different names for the same instruction. For example the following instructions do the same job:

| *Intel* | MOV r1 r2 | (r1) | (r2) |
| *Zilog* | LD r s | (r) | (s) |

In general the Intel documentation is easier to follow.

Anyone using an 8085 or Z80-based system has a good start with the data and descriptions of the last chapter. It therefore seemed

Figure 7.1 The AIM 65 microcomputer system. *(Courtesy: Pelco Electronics)*

appropriate to use a 6500-based computer for this chapter. It is not as powerful as either of the other two processors mentioned, but is the most popular for small personal computers.

The particular system I have chosen to use is the AIM 65. This computer is made by Rockwell and I am much indebted to Pelco Electronics Ltd for the loan of a machine. A photograph of it is shown as Figure 7.1. It includes

- A full keyboard.
- A strip display.
- A small printer.
- A built-in assembler and operating system (ROM).

It will be described in more detail in a later section.

If you are not using the AIM 65 microcomputer you should perform programming exercises as near as possible to the ones described.

It should now be clear to you that this is a very practical chapter. To do the exercises will require that you really understand how a computer works. A number of problems can be avoided no longer. It is essential that you get access to a suitable computer. Any computer that meets the requirements laid down is potentially suitable. Especially if you have never before run a program, you will have many frustrating problems to overcome. Fortunately the benefits and satisfaction will be equally high. At the end of it you will really begin to feel that computers are both controllable and understandable.

7.2 The 6500 instruction set

Architecturally the main difference between the 6500 and the 8085 is that the former does not have any general-purpose data registers. It has just one accumulator. In consequence operations that involve two numbers require that one be in the accumulator and the other in the memory. Both the 8080 and Z80 can perform this operation too. They can also add two numbers already held in registers within the processor.

Keeping as nearly as possible to the spirit of the sub-set of instructions described in Chapter 6, we will limit our attention to the following 6500 operations.

LOAD (immediate)

Form LDA # data
Action

The data part of the instruction is loaded into the accumulator. Thus LDA 56 will cause the number 56 to be loaded into the accumulator. *Note:* The hash sign (#) is used to indicate that it is the immediate addressing mode.

LOAD (address)

Form LDA address
Action (A) ← (M)

The data stored in the memory location defined by the address is copied into the accumulator.

STORE

Form · STA address
Action (M) ← (A)

The data in the accumulator is copied into the memory location defined by the address part of the instruction.

ADD

Form ADC address
Action (A) ← (A) + (M)

The contents of the accumulator are added to the data stored in the address location defined in the address portion of the instruction. The result is stored in the accumulator. In this operation the carry flag is also used. The carry bit is treated as an lsb (least significant bit) and added in.

Carry is a single bit flag. When two numbers are added together the operation may result in a carry. Then the carry flag is set to a 1. Otherwise it is set fo a 0.

Increment

Form   INC address
Action $(M) \leftarrow (M) + 1$

The contents of the specified memory location are increased by one.

Decrement

Form   DEC address
Action $(M) \leftarrow (M) - 1$

The contents of the specified memory location are decreased by one.

No OPeration

Form   NOP
Action   Nothing changes

This is a genuine instruction. It takes a full instruction time to execute, but leaves everything (the machine state) except the program counter just as it was.

Compare (immediate)

Form   CMP data
Action

Two numbers are compared. One is that stored in the accumulator; the other is the data part of the instruction. Both numbers are left unchanged. It is the flags that are affected. In particular the NZ and C flags. By inspecting the flags after a CMP instruction one can deduce if the numbers were identical, and if not which is the larger.

Jump

Form   JMP address
Action $(PC) \leftarrow$ address

This is a normal jump instruction. It causes the address part of the instruction to be loaded into the program counter. In consequence the next instruction will be fetched from that address.

Set and Clear Carry

Form	SEC
	CLC
Action	(C) ← 1 (SEC)
	(C) ← 0 (CLC)

The mnemonics have the meanings:

SEt   Carry flag   (SEC)

CLear   Carry flag   (CLC)

One sets the carry flag, the other *clears it*.

Branch on Carry

Form	BCS displacement
	BCC displacement
Action	IF the condition is met, i.e.
	Carry = 1 for BCS (carry set)
	Carry = 0 for BCC (carry clear)
	THEN
	(PC) ← (PC) + displacement

This is a type of addressing mode not previously described. The 8085 does not have this facility. The Z80 does. The idea of this instruction is that the displacement defines how many bytes of program are to be jumped over.

7.3   Hexadecimal

Before actually starting to run programs there is one further topic that we must cover; hexadecimal coding. It is a system that we use to avoid the need to write and think in binary, unless it is absolutely unavoidable.

The coding scheme is shown in Table 7.1. Once the idea of using letters to represent numbers is accepted the whole scheme makes very good sense. By writing just one symbol we can represent any of the 16 possible sets of 4 bits. With a little practice you will soon visualise that F means a string of ones. With more frequent use you will be able mentally to picture the binary pattern associated with each of the 16 hexadecimal symbols. For brevity we usually say that numbers written in hexadecimal are in *hex*.

A particularly common use of hex is to represent a byte. Then FF means 11111111, i.e. eight ones. In this way all of the 256 binary combinations that can be represented by a byte, are compressed into two symbols; a two-digit hex number. Similarly a 16-bit address can

Table 7.1 The hexadecimal (hex) coding scheme

Decimal	Binary	Hexadecimal
0	0000	0
1	0001	1
2	0010	2
3	0011	3
4	0100	4
5	0101	5
6	0110	6
7	0111	7
8	1000	8
9	1001	9
10	1010	A
11	1011	B
12	1100	C
13	1101	D
14	1110	E
15	1111	F

Table 7.2 Hexadecimal coding of one-byte numbers

Hex	Dec	Hex	Dec
←		1 byte	→
0	0	0	0
1	16	1	1
2	32	2	2
3	48	3	3
4	64	4	4
5	80	5	5
6	96	6	6
7	112	7	7
8	128	8	8
9	144	9	9
A	160	A	10
B	176	B	11
C	192	C	12
D	208	D	13
E	224	E	14
F	240	F	15

Example: 67 (hex) = 96 + 7 = 103 (decimal)

be represented by two bytes and therefore by a 4-digit hex number. Thus FFFF is the hex for 1111111111111111, i.e. sixteen ones.

The hex code is an industry-wide standard. All of the reference manuals for microcomputers use hex in defining the instructions. To take just one example, the 6500 compare immediate instruction is CMP data. Inspection of the reference manual shows the opcode as being C9 in hex.

At this point you should have a look at the reference manual for whatever microprocessor chip you are intending to use. Check through the hex opcodes for a few instructions, until you gain some familiarity with the convention. Try writing down the binary equivalents if you need convincing that it is a good idea.

You should also try converting backwards and forwards from hex to decimal. In doing this you will find the chart of Table 7.2 helpful.

7.4   Trying out the instruction set

Whilst working through this section you should have at your side whichever microcomputer you are going to use. If you have an AIM 65, that will be ideal. If not, your chosen computer will have facilities for doing much the same sort of things. In either case you will need reference manuals. In particular you should have:

- Hardware/technical manual for the processor chip.
- Programming manual for the processor chip.
- User's manual for the microcomputer.
- Any available reference cards.

Now that we are at last ready, put the computer by your elbow and switch on. Now what? Each system will respond in its own individual way. The AIM 65 is particularly friendly. It prints out on its own printer:

```
ROCKWELL AIM 65
```

On the display, it outputs ∠. Now what? The AIM 65 is waiting to be told what we want it to do. We have just a few alternatives, but to find out what they are you must refer to the 'user's guide'. Better still, if you can tolerate its cryptic shortness, is the AIM 65 summary card.

The computer is in the *monitor*. For our first efforts we want the 'I' command. So hit the 'I' key.

To remind us that we *did* the printer responds with

<I>

The display shows

0000∧

Now what? Well the reference card tells us that we have entered a command that allows us to 'enter instructions'. It also tells us;

AAAA [*] = [Address]

What does this mean? Try entering an *, the display responds with

0000 * = ∧

Obviously it wants us to enter an address – not a decimal address, but a hex address. So type in ABCD, a perfectly valid hex address. We now have displayed

0000 * = ABCD

Now what? Try hitting return. Hitting the space bar will do equally well. Computer programs and operating systems are very often written so that the action is completed with a return. The AIM 65 is no exception. It responds in two ways. The display becomes:

A B C D

The printer shows us what we did:

0 0 0 0 * = A B C D

But what have we accomplished? We have defined the address at which the program – yet to be entered – will start.

In fact 0005 is a much better place to start a program. So redefine the address, this time to 0005. But how? This is a good question. It is one that you may have needed answering earlier when a small mistake caused matters to go in the wrong direction. Reversing a wrong direction is often awkward. In fact good facilities exist in any system. At present we will use the simplest and most universal technique. Hit reset. Whenever you get into a mess hit reset.

Now we are back at the switch-on point. So hit 'I'. Notice that we are not in fact back to exactly *where we started*. The computer has remembered the ABCD. Hitting reset does not destroy all the stored data. It just gets the program counter back to zero and allows it to restart at the program that is stored there. Exactly what it then does depends on that program. In our case it enters the monitor, and waits to be told what to do:

*0005

leads to the display of

0005

Now we can enter an instruction. A good one to try first is:

LDA # data

Do not forget the hash sign (#).

We will use the number 23. So enter, by hitting the appropriate keys in sequence,

LDA # 23

Notice that the computer has automatically put a space between the mnemonic and the data. Now hit return. The printer has recorded our action, and the display responds with

0007

Why? Well the instruction we entered had two bytes: one byte of opcode plus one byte of data. It has now occupied two adjacent memory locations. It started at 0005, has taken two spaces and the next instruction must start at 0007.

For simplicity we will follow the last instruction with

LDA # 25

So proceed as before. If you make a mistake before hitting return, the DEL key will let you make deletions; provided you do not hit return until corrected.

Now we will use the store instruction.

STA address

In a similar way enter

STA 0010
NOP
BRK

Only the last of these is unfamiliar to you. It is not a real instruction, but has an important effect. It will cause the entry of a rather special instruction, one that we are not concerned with understanding. Its effect is to alter the flow of control. The result is that BRK defines the end of the program *and* returns control to the monitor. If the program is not terminated in this way there is no knowing what will happen next. Moreover, the monitor would not be in control, the keyboard would be ignored. Only by the hitting of reset could your wishes be made known. The results of running your program would probably have disappeared, due to overwriting.

We are now ready to enter another part of the monitor. We want to run the program, and to see what happens. We will start with the M command. This lets us display and alter the memory contents.

First let us look at the printer output. It tells us what we will expect to find in the memory locations into which we have entered our program. The roll shows:

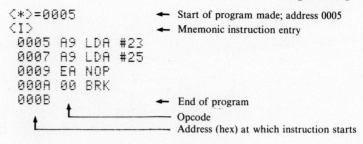

```
<*>=0005                    ← Start of program made; address 0005
<I>                         ← Mnemonic instruction entry
 0005 A9 LDA #23
 0007 A9 LDA #25
 0009 EA NOP
 000A 00 BRK
 000B                       ← End of program
                            ── Opcode
                            ── Address (hex) at which instruction starts
```

Study this list. It shows us that the memory locations, starting at 0005, are now containing hex data, as follows:

Address	Data (in hex)
0005	A9
	23
0007	A9
	25
0009	EA
000A	00

That this is correct can be more clearly understood by using the AIM 65 in a slightly different way. Go back and re-enter the program, but use the space bar instead of the return. To start this off hit ESC (escape). This will return the computer to the monitor.

When you have re-entered the program the printer will show

```
<
<*>=0005
<I>
 0005 A9 LDA #23
 0005 A9 23
 0007 A9 LDA #25    ⎤ Data part of instruction
 0007 A9 25         ⎦
 0009 EA NOP
 000A 00 BRK
 000B
<                  ──────── Opcodes
```

Now re-enter the monitor with ESC. Then hit M. This enters the program for 'Display/Alter Memory Contents'. The display responds with

$$(M) = \wedge$$

The computer wants to know the address at which we wish to start examining the memory contents. We want to start at the beginning of the program, i.e. at 0005. So we enter this number. Now hit the space bar. On the printer, we are presented a display with

    0005   A9 23 A9 25

Now hit the space bar. We now get

0009   EA 00 EA EA

This is comforting. It is what we were led to expect. The actual print-out is:

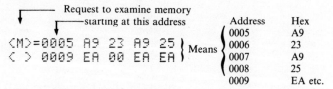

The contents of 000A onwards are of no significance.

Regardless of which computer you are using you should achieve this level of progress before proceeding. All we have done is:

- Entered a sequence of just a few simple instructions.
- Obtained the hex code equivalent.
- Looked at memory, to check that the program is correct.

Now we will run the program, and see what it does.

- Return to the monitor – use ESC.
- Go into single step – set the appropriate switch.

To run our program, which starts at address 0005, we must set the program counter to that number. We can do this by using the * command. As previously the display responds with

$$< * > = \wedge$$

We type in 0005 and terminate with space.

We are automatically returned to the monitor. Now hit G. This is the GO command. Now enter a RETURN. We are greeted by a display reading 0007 A5 LDA # 25. This as we know is the *next* instruction. LDA # 23 has already been executed.

Now we will have a look at the accumulator. Since we have executed LDA #23, it should contain 23. So hit R. This is the 'examine registers command'. We are greeted by a display reading

0007 20 23 00 FF

The printer shows:

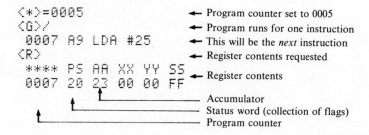

Notice the headings PS AA XX YY SS. They refer to five registers. AA is the accumulator. It is shown as holding 23, as expected.

Now execute the next instruction with another G command, and re-examine the registers with another R command. Printer and display show all the registers unchanged, except the accumulator, which now holds 25. Now complete the program by using two or more pairs of G then R commands. The print-out is now:

```
<*>=0005                    Program counter set to 0005
<G>/                        First instruction executed
 0007 A9 LDA #25            This will be next
<R>
 **** PS AA XX YY SS    }   Registers examined
 0007 20 23 00 00 FF
<G>/
 0009 EA NOP
<R>                         Next instruction executed
 **** PS AA XX YY SS        Results examined
 0009 20 25 00 00 FF
<G>/
 000A 00 BRK
<R>                         Ditto
 **** PS AA XX YY SS
 000A 20 25 00 00 FF
<G>/
 000C EA NOP
<R>                         Ditto
 **** PS AA XX YY SS
 000C 30 25 00 00 FF
```

Notice that NOP and BRK changed nothing.

Once you have run the program described, or a similar one, you have surmounted the major hurdle to using a computer. You will probably have noticed how on the one hand the computer is absolutely unforgiving in terms of errors. You must get everything exactly right. On the more positive side you will also now be aware how obedient the computer is. No damage whatsoever is done by making experiments. If you are not sure how an instruction works you can try it and see what happens.

Before writing a program to do something useful you will do well to become familiar with the instruction set. If you are using the AIM 65 or another 6500-based computer you should try out all the instructions described in Section 7.2. In the event that you are using a Z80 or 8085 you should familiarise yourself with the instructions described in Chapter 6.

With the AIM 65 you could try the following program sequence:

```
*0005                    Starts program at address 0005
I                        Enters mnemonic entry mode
LDA     #23              Puts 23 into accumulator
STA     0050             Loads address 0050 from accumulator
INC     0050             Increments (0050)
LDA     0050             Loads accumulator from 0050
ADC     0050             Adds (0050) to accumulator
SEC                      Sets Carry (to 1)
CLC                      Clears Carry (to 0)
CMP     47               Compare accumulator to 47
CMP     48               Compare accumulator to 48
CMP     49               Compare accumulator to 49
JMP     0007             Jumps to instruction at 0007 (STA 0050)
```

The results of entering this program, and then running it, step by step, are shown below. Note the inspection of registers after each step. There is also a memory inspection. To do this use the M command.

The computer responds with a request for a definition of the address to be looked at, 0050 in our case. A RETURN leads to the display of the memory contents.

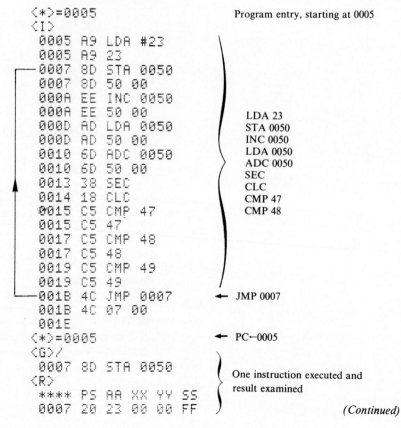

```
<*>=0005                        Program entry, starting at 0005
<I>
 0005 A9 LDA #23
 0005 A9 23
 0007 8D STA 0050
 0007 8D 50 00
 000A EE INC 0050
 000A EE 50 00
 000D AD LDA 0050           LDA 23
 000D AD 50 00              STA 0050
 0010 6D ADC 0050           INC 0050
 0010 6D 50 00              LDA 0050
 0013 38 SEC                ADC 0050
 0014 18 CLC                SEC
 0015 C5 CMP 47             CLC
 0015 C5 47                 CMP 47
 0017 C5 CMP 48             CMP 48
 0017 C5 48
 0019 C5 CMP 49
 0019 C5 49
 001B 4C JMP 0007        ← JMP 0007
 001B 4C 07 00
 001E
<*>=0005                 ← PC←0005
<G>/
 0007 8D STA 0050
<R>                         One instruction executed and
 **** PS AA XX YY SS        result examined
 0007 20 23 00 00 FF                              (Continued)
```

```
<G>/
 000A EE INC 0050
<R>
 **** PS AA XX YY SS                    Ditto
 000A 20 23 00 00 FF
<M>=0050 23 D8 80 45
<G>/
 000D AD LDA 0050
<R>
 **** PS AA XX YY SS                    Ditto
 000D 20 23 00 00 FF
<M>=0050 24 D8 80 45
<G>/
 0010 6D ADC 0050
<R>
 **** PS AA XX YY SS                    Ditto
 0010 20 24 00 00 FF
<G>/
 0013 38 SEC
<R>                                     Ditto
 **** PS AA XX YY SS
 0013 20 48 00 00 FF
<G>/
 0014 18 CLC
<R>                                     Ditto
 **** PS AA XX YY S9
 0014 21 48 00 00 FF
<G>/
 0015 C5 CMP 47
<R>                                     Ditto
 **** PS AA XX YY SS
 0015 20 48 00 00 FF
<G>/
 0017 C5 CMP 48
<R>                                     Ditto
 **** PS AA XX YY SS
 0017 20 48 00 00 FF
<G>/
 0019 C5 CMP 49
<R>                                     Ditto
 **** PS AA XX YY SS
 0019 21 48 00 00 FF
<G>/
 001B 4C JMP 0007
<R>                                     Ditto
 **** PS AA XX YY SS                    Memory examined also
 001B 20 48 00 00 FF                    Note next instruction is a jump
<M>=0050 24 D8 80 45
```

(Continued)

```
<G>/
 0007 8D STA 0050
<R>
 **** PS AA XX YY SS
 0007 20 48 00 00 FF
<M>=0050 24 D8 80 45
<G>/
 000A EE INC 0050
<R>
 **** PS AA XX YY SS
 000A 20 48 00 00 FF
<M>=0050 48 D8 80 45
<G>/
 000D AD LDA 0050
<R>
 **** PS AA XX YY SS
 000D 20 48 00 00 FF
<M>=0050 49 D8 80 45
<G>/
 0010 6D ADC 0050
<R>
 **** PS AA XX YY SS
 0010 20 49 00 00 FF
<M>=0050 49 D8 80 45
<G>/
 0013 38 SEC
<R>
 **** PS AA XX YY SS
 0013 E0 92 00 00 FF
<M>=0050 49 D8 80 45
<R>
 **** PS AA XX YY SS
 0013 E0 92 00 00 FF
<G>/
 0014 18 CLC
<R>
 **** PS AA XX YY SS
 0014 E1 92 00 00 FF
<M>=0050 49 D8 80 45
<G>/
 0015 C5 CMP 47
<R>
 **** PS AA XX YY SS
 0015 E0 92 00 00 FF
<G>/
 0017 C5 CMP 48
<R>
 **** PS AA XX YY SS
 0017 E0 92 00 00 FF
<G>/
 0019 C5 CMP 49
<R>
 **** PS AA XX YY SS
```

Ditto
Note PC now 0007

As before
Note numbers are different this time

Etc.   Etc.

(Continued)

```
 0019 61 92 00 00 FF
<G>/
 001B 4C JMP 0007
<R>
 **** PS AA XX YY SS
 001B E0 92 00 00 FF
<001B E0 92 00 00 FF
```

In this example there is a loop generated with the JMP instruction. Notice how the increment and add operation causes the numbers to increase each time the loop is traversed. In the first run-through the original 23 is incremented to 24 and then added to itself to give 48. In the second traverse we start with 48 instead of 23.

When the addition command is executed for the second time it is 49 and 49 that are added together. The resulting answer is 92. Wrong you may think. In fact the answer given by the computer is correct. The explanation lies in the nature of the numbers. They are hexadecimal and not decimal. When 9 and 9 are added we expect the answer 18. Now this is equal to 16 plus 2. Sixteen generates a carry and leaves 2. Thus 49 (hex) + 49 (hex) = 92 (hex) (see Table 7.2).

An alternative viewpoint is

$$49 \text{ (hex) to.} \quad (4 \times 16) + 9 = \quad 64 + 9 = \quad 73$$
$$49 \text{ (hex) to.} \quad (4 \times 16) + 9 = \quad 64 + 9 = \quad 73$$
$$92 \text{ (hex) to.} \quad (9 \times 16) + 2 = 144 + 2 = 146$$

Do not proceed to the next section until you have tried out all the instructions that have been discussed.

7.5 Writing and running programs

Now that you can drive your computer and understand a small group of instructions, you are ready to expand your horizons. In this section will be provided a framework for systematically extending your knowledge and experience.

First we will write a more general-purpose program. Our example is a short program that adds two numbers; the numbers that are to be added we will call A and B. The answer, A + B, we will call C. Each of these numbers is to be a single byte. We will represent everything in hex. The inputs A and B are to be placed in two specific memory locations, and the answer is to be returned to a third.

We proceed by allocating a specific storage location to each number. I have chosen to make the following allocation.

A is to be at address 0060 (hex)
B is to be at address 0061 (hex)
C is to be at address 0062 (hex)

You may use any locations you choose. The restrictions are these:

- There must be RAM in the locations used.
- The chosen locations must not be used by:
  - (a) the program;
  - (b) the system software.

To add A and B to produce C, a suitable program would be:

```
*                  = 0005
<I>                To define start of program
CLC                Clear carry
LDA      0060      A to accumulator
ADC      0061      (A+B) to accumulator
STA      0062      (A+B) to C
BRK                End
NOP                Leaves a 'clean' end
NOP
```

To run the program we must first place A and B in location 0060 and 0061. To do this we can use the M command. I chose to enter 02 and 03. The result of the whole process, including looking at 0062 for the answer, is shown below.

```
<*>=0005                          ◄── Start program at 0005
<I>
 0005 EA NOP                                              ⎫
 0006 18 CLC                                              ⎪
 0007 AD LDA 0060    ⎫                                    ⎬ Program
 000A 6D ADC 0061    ⎬ A, B and C used                   ⎪
 000D 8D STA 0062    ⎭                                    ⎪
 0010 00 BRK                                              ⎪
 0011 EA NOP         ⎫ NOPs give                          ⎪
 0012 EA NOP         ⎭ 'clean' end                        ⎭
 0013
<
<M>=0060  00 00 00 0              ◄── Memory examined
0
</> 0060 02 03                    ◄── A set to 02, B to 03
<*>=0005                          ◄── PC←0005
<G>/
 0006 18 CLC                                              ⎫
<G>/                                                      ⎪
 0007 AD LDA 0060                                         ⎪
<G>/                                                      ⎬ Program single-stepped
 000A 6D ADC 0061                                         ⎪
<G>/                                                      ⎪
 000D 8D STA 0062                                         ⎭
```

(Continued opposite)

```
<G>/
  0010 00 BRK
<G>/
  0012 EA NOP
<0012 EA NOP
<
<M>=0060 02 03 05 00        Memory examined; result,
<M>=0060 02 03 05 00        02 + 03 = 05 stored in 0062
```

Now we will run the program straight through, without single stepping. To do this on the AIM 65 move the appropriate switch from single step to run. Now when we tell the program to GO, it will run to completion. It will re-enter the monitor at BRK.

First we must enter into the A and B locations the two numbers we wish to add. I entered 06 and 0F. They are the equivalent to decimal 6 and 15. Thus we will expect as an answer the decimal 21 which is 15 (hex). The results of doing this are:

```
<M>=0060 51 2C 88 DD  ← Memory contents displayed 0060↑
</> 0060 06 0F 00 00  ← A set to 06, B to 0F, C to 00
<*>=0005
<G>/                       Program runs to completion
  0011 EA NOP
<M>=0060 06 0F 15 00  ← Memory examined
<R>
 **** PS AA XX YY SS
 0011 30 15 00 00 FF  ← Registers examined
```

Now you should try several examples. In particular try adding two numbers that give a carry bit. For example 0F + F2 = ? Try it. The results of my session were:

```
<M>=0060 00 00 00 00        A to 0F, B to F2
</> 0060 0F F2 00 00
<*>=0005
<G>/                        Program runs to completion
  0011 EA NOP
<0011 EA NOP
<M>=0060 0F F2 01 00
<R>
 **** PS AA XX YY SS        Results examined; status word
 0011 31 01 00 00 FF        has changed
<0011 31 01 00 00 FF
<
```

Notice that the status word is 31. Previously it was 30. Thus the status word changed from

00110000

to

00110001

The change is because of the carry. It was to avoid the carry bit causing problems that the CLC instruction was included in the program. Had it not been there the result of the ADC instruction would have been one higher, if the carry bit were a 1 when the program was executed. You can easily try out this feature by replacing the CLC with SEC, which sets the carry flag to a 1. Then 02 plus 03 becomes 06 because the instruction is really add with carry. This feature enables us to add larger numbers, the carry flag acting as a link to the next higher byte.

Next we will look at the idea of the program having conditional tests in it. Then we can create logic of the form

IF condition x   perform task A
ELSE   perform task B

A flowchart for this type of algorithm is shown as Figure 7.2. There are many examples of programs having this sort of structure. For instance a payroll program might include a test to see if overtime had been done. A process control computer trying to stabilise a temperature might be programmed to see if the temperature was too high or too low. Depending on the result of this test it would be required to run either the heating program or the cooling program.

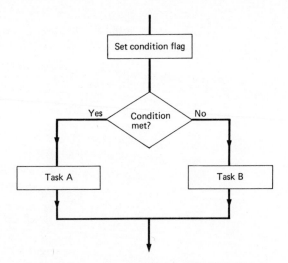

Figure 7.2   Flowchart for IF condition met do A ELSE do B

A suitable program to try out this idea is:

```
*       0005
I       NOP
        LDA     0060
        ADC     0061
        STA     0062
        BCC     0019    Jmp 3 instr. if carry = 0
        LDA     #FF     FF to 0063
        STA     0063    If carry = 1
        JMP     001E    Go to end
        LDA     #AA     AA to 0063
        STA     0063    If carry = 0
        BRK
        NOP
        NOP
```

The general idea is to add two numbers and to put either FF or AA into location 0063, depending on the state of the carry flag after the addition.

In entering this program two problems arise. Both the JMP and BCC instructions need an address. To put in the correct numbers requires you to discover the address of the instruction to which they must jump. One way is to take a guess and enter the program. Then look at the listing to see what the jump addresses ought to be. You can then make corrections to just the relevant instructions. To do this you use the * command to define the address at which you wish to change something. After an I command you can re-enter the instruction, with the corrected address. This is the easiest method, and the one I used to discover the correct addresses.

An alternative method is to calculate the offset. For instance the BCC instruction has to jump over.

```
    LDA # FF    – 2 bytes
    STA 0063    – 3 bytes
    JMP 001E    – 3 bytes
    A total of  – 8 bytes
```

With the I code entry this is achieved by entering the code

    BCC 08

Reference to the coding sheet will reveal that this is the appropriate entry for a *forward jump* over 08 bytes. The computer adds this to the value that the program counter would be at at the start of the next instruction, adds the two numbers together and prints out and stores the code BCC 0019. So on execution the program branches to the instruction at 0019 *if* the carry flag is clear, i.e. is a 0. The results of running this program are:

```
                        <*>=0005
                        <I>
                         0005 EA NOP
                         0006 AD LDA 0060
                         0009 6D ADC 0061
                         000C 8D STA 0062
                         000F 90 BCC 0019        Program starting at 0005
                         0011 A9 LDA #FF         runs only if C = 1
             IF c = 0    0013 8D STA 0063
                         0016 4C JMP 001E
                         0019 A9 LDA #AA
                         001B 8D STA 0063
                         001E 00 BRK
                         001F EA NOP
                         0020 EA NOP
                         0021                    END
                        <
                        <
                        <M>=0060  03 04 08 0     0060←04
                        0                        0061←05
                        </> 0060 04 05 00 00     0062←00
                        <*>=0005                 0063←00
                        <G>/
                         001F EA NOP             Program runs
                        <R>
                         **** PS AA XX YY SS
                         001F B0 AA 00 00 FF     Results examined
                        <M>=0060 04 05 09 AA     0062 = 09; 0063 = AA [C = 0]
                        <M>=0060 04 05 09 AA
                        <
                        <
                        <M>=0060 04 05 09 AA     0060←0F
                        </> 0060 0F F2 00 00     0061←F2
                        <*>=0005                 0062←00
                        <G>/                     0063←00
                         001F EA NOP
                        <R>                      Program runs
                         **** PS AA XX YY SS
                         001F B1 FF 00 00 FF     Results examined
                        <M>=0060 0F F2 01 FF     0062 = 01; 0063 = FF [C = 1]
                        <M>=0060 0F F2 01 FF
                        <
```

Finally we will write a program that loops. A program with this feature was introduced in the last section, but it had an impractical feature. The loop was never-ending.

Practical programs with loops must include a terminating condition. The general idea is shown in the flow diagram of Figure 7.3. We will simplify as far as possible; we will not bother much with the program

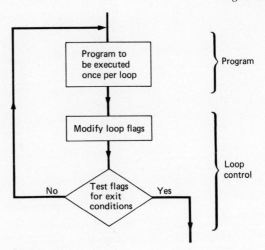

Figure 7.3 Flowchart for a loop

part within the loop. Instead we will concentrate on making the computer go round the loop a defined number of times. We will use the following algorithm.

Enter a number to memory location A
Clear Carry Flag
Add 1 to A
Loop if carry = 0
BRK

The result of doing this is shown below:

```
<*>=0005
<I>
 0005 EA NOP
 0006 18 CLC
 0007 A9 LDA #01
 0009 6D ADC 0060
 000C 8D STA 0060
 000F 90 BCC 0007
 0011 00 BRK
 0012 EA NOP
 0013 EA NOP
 0014
<
<
<M>=0060  88 00 00 0
0
</> 0060 F0 00 00 00
```

IF C = 0

} 1 added to (0060)

↓ IF C = 1
Note FF + 01 generates a carry
i.e. FF + 01 = 00 and C = 1

← (0060)←F0

(Continued overleaf)

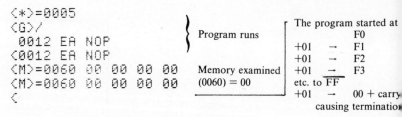

```
<*>=0005
<G>/                          ⎫
  0012 EA NOP                 ⎬  Program runs
<0012 EA NOP                  ⎭
<M>=0060 00 00 00 00             Memory examined
<M>=0060 00 00 00 00             (0060) = 00
<
```

	The program started at F0
+01 →	F1
+01 →	F2
+01 →	F3
etc. to F̄F̄	
+01 →	00 + carry
	causing termination

Notice that the number entered has become zero. The program kept adding 1 until the number got to FF, the next time round there would be a carry generated and the loop terminated.

During the writing and running of your test programs you will inevitably have met problems that had to be overcome. You may for instance have run a program that never ended! With the AIM 65, for instance, the display blanks-out and the computer appears dead. In reality it is busy executing a never-ending program. A reset will restore control.

We have now come, virtually, to the end of this first course on microelectronics. Hopefully you will be well prepared for more advanced courses. You have covered all of the basic ideas that you need to grapple with both microelectronic hardware and computer software. Moreover you should have gained sufficient confidence to start to gain experience by experiment. One of the wonderful things about computers is that you can make endless experiments, learning more and more; a computer can be your willing and tireless friend.

Exercises

7.1 For all of the instructions that we have covered, and a few more as well, produce a chart showing the hex codes for the instructions. You can discover the hex codes by studying the manufacturers' literature.

7.2 For all the instructions studied in Exercise 7.1 write down what it does. Check that you have properly understood its action by using your computer.

7.3 Write a short program and translate it to hex by hand, i.e. use pencil, paper and reference books. Then assemble it on your computer, and check that you did it correctly.

7.4 Write, run and test:
  (a) A program that adds together three numbers.
  (b) A looping program that gives a result equal to $5n$, where n is the number of times round the loop.

Appendix 1 Data on the Intel 8085 microprocessor

The following pages are reproduced, by kind permission of Intel Corporation, direct from the Intel Data Book of the 8085 microprocessor. They give an overview, a minimum system, a description of the instruction set and a summary of the instruction set.

8085

SINGLE CHIP 8-BIT N-CHANNEL MICROPROCESSOR

- Single +5V Power Supply
- 100% Software Compatible with 8080A
- 1.3 μs Instruction Cycle
- On-Chip Clock Generator (with External Crystal or RC Network)
- On-Chip System Controller

- Four Vectored Interrupts (One is non-Maskable)
- Serial In/Serial Out Port
- Decimal, Binary and Double Precision Arithmetic
- Direct Addressing Capability to 64K Bytes of Memory

The Intel® 8085 is a new generation, complete 8 bit parallel central processing unit (CPU). Its instruction set is 100% software compatible with the 8080A microprocessor, and it is designed to improve the present 8080's performance by higher system speed. Its high level of system integration allows a minimum system of three IC's: 8085 (CPU), 8155 (RAM) and 8355/8755 (ROM/PROM).

The 8085 incorporates all of the features that the 8224 (clock generator) and 8228 (system controller) provided for the 8080, thereby offering a high level of system integration.

The 8085 uses a multiplexed Data Bus. The address is split between the 8 bit address bus and the 8 bit data bus. The on-chip address latches of 8155/8355/8755 memory products allows a direct interface with 8085.

8085 CPU FUNCTIONAL BLOCK DIAGRAM

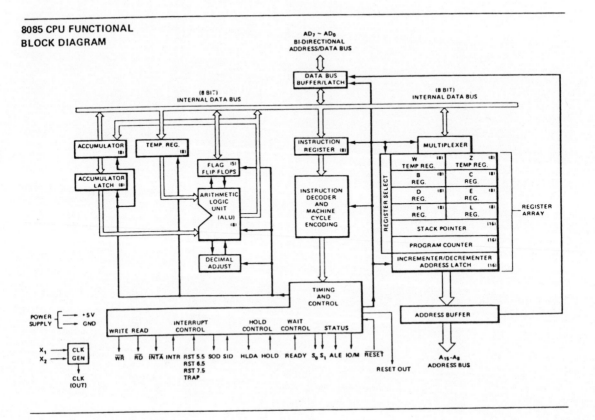

8085

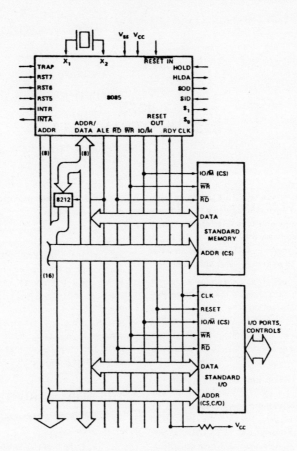

FIGURE 6. MCS-85™ MINIMUM SYSTEM (USING STANDARD MEMORIES).

8085
INSTRUCTION SET*

A computer, no matter how sophisticated, can only do what it is "told" to do. One "tells" the computer what to do via a series of coded instructions referred to as a Program. The realm of the programmer is referred to as Software, in contrast to the Hardware that comprises the actual computer equipment. A computer's software refers to all of the programs that have been written for that computer.

When a computer is designed, the engineers provide the Central Processing Unit (CPU) with the ability to perform a particular set of operations. The CPU is designed such that a specific operation is performed when the CPU control logic decodes a particular instruction. Consequently, the operations that can be performed by a CPU define the computer's Instruction Set.

Each computer instruction allows the programmer to initiate the performance of a specific operation. All computers implement certain arithmetic operations in their instruction set, such as an instruction to add the contents of two registers. Often logical operations (e.g., OR the contents of two registers) and register operate instructions (e.g., increment a register) are included in the instruction set. A computer's instruction set will also have instructions that move data between registers, between a register and memory, and between a register and an I/O device. Most instruction sets also provide Conditional Instructions. A conditional instruction specifies an operation to be performed only if certain conditions have been met; for example, jump to a particular instruction if the result of the last operation was zero. Conditional instructions provide a program with a decision-making capability.

By logically organizing a sequence of instructions into a coherent program, the programmer can "tell" the computer to perform a very specific and useful function.

The computer, however, can only execute programs whose instructions are in a binary coded form (i.e., a series of 1's and 0's), that is called Machine Code. Because it would be extremely cumbersome to program in machine code, programming languages have been developed. There are programs available which convert the programming language instructions into machine code that can be interpreted by the processor.

One type of programming language is Assembly Language. A unique assembly language mnemonic is assigned to each of the computer's instructions. The programmer can write a program (called the Source Program) using these mnemonics and certain operands; the source program is then converted into machine instructions (called the Object Code). Each assembly language instruction is converted into one machine code instruction (1 or more bytes) by an Assembler program. Assembly languages are usually machine dependent (i.e., they are usually able to run on only one type of computer).

THE 8085 INSTRUCTION SET

The 8085 instruction set includes five different types of instructions:

- **Data Transfer Group**—move data between registers or between memory and registers

- **Arithmetic Group** — add, subtract, increment or decrement data in registers or in memory

- **Logical Group** — AND, OR, EXCLUSIVE-OR, compare, rotate or complement data in registers or in memory

- **Branch Group** — conditional and unconditional jump instructions, subroutine call instructions and return instructions

- **Stack, I/O and Machine Control Group** — includes I/O instructions, as well as instructions for maintaining the stack and internal control flags.

Instruction and Data Formats:

Memory for the 8085 is organized into 8-bit quantities, called Bytes. Each byte has a unique 16-bit binary address corresponding to its sequential position in memory.

The 8085 can directly address up to 65,536 bytes of memory, which may consist of both read-only memory (ROM) elements and random-access memory (RAM) elements (read/write memory).

Data in the 8085 is stored in the form of 8-bit binary integers:

DATA WORD

D_7	D_6	D_5	D_4	D_3	D_2	D_1	D_0

MSB   LSB

When a register or data word contains a binary number, it is necessary to establish the order in which the bits of the number are written. In the Intel 8085, BIT 0 is referred to as the **Least Significant Bit (LSB)**, and BIT 7 (of an 8 bit number) is referred to as the **Most Significant Bit (MSB)**.

The 8085 program instructions may be one, two or three bytes in length. Multiple byte instructions must be stored in successive memory locations; the address of the first byte is always used as the address of the instructions. The exact instruction format will depend on the particular operation to be executed.

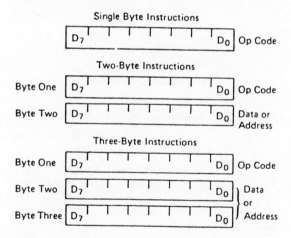

Single Byte Instructions

D_7 .... D_0 Op Code

Two-Byte Instructions

Byte One D_7 .... D_0 Op Code

Byte Two D_7 .... D_0 Data or Address

Three-Byte Instructions

Byte One D_7 .... D_0 Op Code

Byte Two D_7 .... D_0 } Data

Byte Three D_7 .... D_0 } or Address

Addressing Modes:

Often the data that is to be operated on is stored in memory. When multi-byte numeric data is used, the data, like instructions, is stored in successive memory locations, with the least significant byte first, followed by increasingly significant bytes. The 8085 has four different modes for addressing data stored in memory or in registers:

- Direct — Bytes 2 and 3 of the instruction contain the exact memory address of the data item (the low-order bits of the address are in byte 2, the high-order bits in byte 3).
- Register — The instruction specifies the register or register-pair in which the data is located.
- Register Indirect — The instruction specifies a register-pair which contains the memory address where the data is located (the high-order bits of the address are in the first register of the pair, the low-order bits in the second).
- Immediate — The instruction contains the data itself. This is either an 8-bit quantity or a 16-bit quantity (least significant byte first, most significant byte second).

Unless directed by an interrupt or branch instruction, the execution of instructions proceeds through consecutively increasing memory locations. A branch instruction can specify the address of the next instruction to be executed in one of two ways:

- Direct — The branch instruction contains the address of the next instruction to be executed. (Except for the 'RST' instruction, byte 2 contains the low-order address and byte 3 the high-order address.)
- Register indirect — The branch instruction indicates a register-pair which contains the address of the next instruction to be executed. (The high-order bits of the address are in the first register of the pair, the low-order bits in the second.)

The RST instruction is a special one-byte call instruction (usually used during interrupt sequences). RST includes a three-bit field; program control is transferred to the instruction whose address is eight times the contents of this three-bit field.

Condition Flags:

There are five condition flags associated with the execution of instructions on the 8085. They are Zero, Sign, Parity, Carry, and Auxiliary Carry, and are each represented by a 1-bit register in the CPU. A flag is "set" by forcing the bit to 1, "reset" by forcing the bit to 0.

Unless indicated otherwise, when an instruction affects a flag, it affects it in the following manner:

Zero: If the result of an instruction has the value 0, this flag is set; otherwise it is reset.

Sign: If the most significant bit of the result of the operation has the value 1, this flag is set; otherwise it is reset.

Parity: If the modulo 2 sum of the bits of the result of the operation is 0, (i.e., if the result has even parity), this flag is set; otherwise it is reset (i.e., if the result has odd parity).

Carry: If the instruction resulted in a carry (from addition), or a borrow (from subtraction or a comparison) out of the high-order bit, this flag is set; otherwise it is reset.

Auxiliary Carry: If the instruction caused a carry out of bit 3 and into bit 4 of the resulting value, the auxiliary carry is set; otherwise it is reset. This flag is affected by single precision additions, subtractions, increments, decrements, comparisons, and logical operations, but is principally used with additions and increments preceding a DAA (Decimal Adjust Accumulator) instruction.

Symbols and Abbreviations:

The following symbols and abbreviations are used in the subsequent description of the 8085 instructions:

SYMBOLS	MEANING
accumulator	Register A
addr	16-bit address quantity
data	8-bit data quantity
data 16	16-bit data quantity
byte 2	The second byte of the instruction
byte 3	The third byte of the instruction
port	8-bit address of an I/O device
r,r1,r2	One of the registers A,B,C,D,E,H,L
DDD,SSS	The bit pattern designating one of the registers A,B,C,D,E,H,L (DDD=destination, SSS= source):

DDD or SSS	REGISTER NAME
111	A
000	B
001	C
010	D
011	E
100	H
101	L

rp   One of the register pairs:

B represents the B,C pair with B as the high-order register and C as the low-order register;

D represents the D,E pair with D as the high-order register and E as the low-order register;

H represents the H,L pair with H as the high-order register and L as the low-order register;

SP represents the 16-bit stack pointer register.

RP   The bit pattern designating one of the register pairs B,D,H,SP:

RP	REGISTER PAIR
00	B-C
01	D-E
10	H-L
11	SP

rh   The first (high-order) register of a designated register pair.

rl   The second (low-order) register of a designated register pair.

PC   16-bit program counter register (PCH and PCL are used to refer to the high-order and low-order 8 bits respectively).

SP   16-bit stack pointer register (SPH and SPL are used to refer to the high-order and low-order 8 bits respectively).

r_m Bit m of the register r (bits are number 7 through 0 from left to right).

Z,S,P,CY,AC The condition flags:

Zero,

Sign,

Parity,

Carry,

and Auxiliary Carry, respectively.

( )   The contents of the memory location or registers enclosed in the parentheses.

◄—   "Is transferred to"

∧   Logical AND

∀   Exclusive OR

∨   Inclusive OR

+   Addition

—   Two's complement subtraction

*   Multiplication

◄—►   "Is exchanged with"

—   The one's complement (e.g., $\overline{(A)}$)

n   The restart number 0 through 7

NNN   The binary representation 000 through 111 for restart number 0 through 7 respectively.

Description Format:

The following pages provide a detailed description of the instruction set of the 8085. Each instruction is described in the following manner:

1. The MCS 85™ macro assembler format, consisting of the instruction mnemonic and operand fields, is printed in **BOLDFACE** on the left side of the first line.

2. The name of the instruction is enclosed in parenthesis on the right side of the first line.

3. The next line(s) contain a symbolic description of the operation of the instruction.

4. This is followed by a narative description of the operation of the instruction.

5. The following line(s) contain the binary fields and patterns that comprise the machine instruction.

6. The last four lines contain incidental information, about the execution of the instruction. The number of machine cycles and states required to execute the instruction are listed first. If the instruction has two possible execution times, as in a Conditional Jump, both times will be listed, separated by a slash. Next, any significant data addressing modes (see Page 4-2) are listed. The last line lists any of the five Flags that are affected by the execution of the instruction.

Data Transfer Group:

This group of instructions transfers data to and from registers and memory. **Condition flags are not affected** by any instruction in this group.

MOV r1, r2   (Move Register)

(r1) ◄— (r2)

The content of register r2 is moved to register r1.

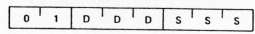

	Cycles:	1
	States:	4
	Addressing:	register
	Flags:	none

MOV r, M   (Move from memory)

(r) ◄— ((H) (L))

The content of the memory location, whose address is in registers H and L, is moved to register r.

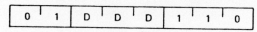

	Cycles:	2
	States:	7
	Addressing:	reg. indirect
	Flags:	none

MOV M, r   (Move to memory)

((H) (L)) ◄— (r)

The content of register r is moved to the memory location whose address is in registers H and L.

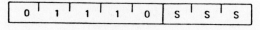

	Cycles:	2
	States:	7
	Addressing:	reg. indirect
	Flags:	none

MVI r, data   (Move Immediate)

(r) ◄— (byte 2)

The content of byte 2 of the instruction is moved to register r.

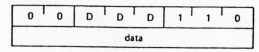

	Cycles:	2
	States:	7
	Addressing:	immediate
	Flags:	none

MVI M, data   (Move to memory immediate)

((H) (L)) ◄— (byte 2)

The content of byte 2 of the instruction is moved to the memory location whose address is in registers H and L.

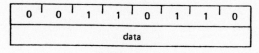

	Cycles:	3
	States:	10
	Addressing:	immed./reg. indirect
	Flags:	none

LXI rp, data 16   (Load register pair immediate)

(rh) ◄— (byte 3),

(rl) ◄— (byte 2)

Byte 3 of the instruction is moved into the high-order register (rh) of the register pair rp. Byte 2 of the instruction is moved into the low-order register (rl) of the register pair rp.

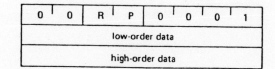

	Cycles:	3
	States:	10
	Addressing:	immediate
	Flags:	none

LDA addr (Load Accumulator direct)

(A) ◄— ((byte 3)(byte 2))

The content of the memory location, whose address is specified in byte 2 and byte 3 of the instruction, is moved to register A.

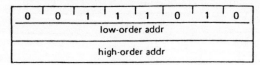

0	0	1	1	1	0	1	0
low-order addr							
high-order addr							

Cycles: 4
States: 13
Addressing: direct
Flags: none

STA addr (Store Accumulator direct)

((byte 3)(byte 2)) ◄— (A)

The content of the accumulator is moved to the memory location whose address is specified in byte 2 and byte 3 of the instruction.

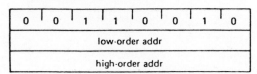

0	0	1	1	0	0	1	0
low-order addr							
high-order addr							

Cycles: 4
States: 13
Addressing: direct
Flags: none

LHLD addr (Load H and L direct)

(L) ◄— ((byte 3)(byte 2))

(H) ◄— ((byte 3)(byte 2) + 1)

The content of the memory location, whose address is specified in byte 2 and byte 3 of the instruction, is moved to register L. The content of the memory location at the succeeding address is moved to register H.

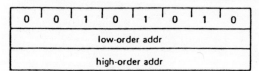

0	0	1	0	1	0	1	0
low-order addr							
high-order addr							

Cycles: 5
States: 16
Addressing: direct
Flags: none

SHLD addr (Store H and L direct)

((byte 3)(byte 2)) ◄— (L)

((byte 3)(byte 2) + 1) ◄— (H)

The content of register L is moved to the memory location whose address is specified in byte 2 and byte 3. The content of register H is moved to the succeeding memory location.

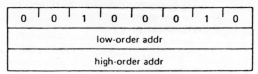

0	0	1	0	0	0	1	0
low-order addr							
high-order addr							

Cycles: 5
States: 16
Addressing: direct
Flags: none

LDAX rp (Load accumulator indirect)

(A) ◄— ((rp))

The content of the memory location, whose address is in the register pair rp, is moved to register A. Note: only register pairs rp=B (registers B and C) or rp=D (registers D and E) may be specified.

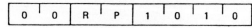

0	0	R	P	1	0	1	0

Cycles: 2
States: 7
Addressing: reg. indirect
Flags: none

STAX rp (Store accumulator indirect)

((rp)) ◄— (A)

The content of register A is moved to the memory location whose address is in the register pair rp. Note: only register pairs rp=B (registers B and C) or rp=D (registers D and E) may be specified.

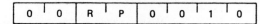

0	0	R	P	0	0	1	0

Cycles: 2
States: 7
Addressing: reg. indirect
Flags: none

XCHG (Exchange H and L with D and E)

(H) ◄—► (D)

(L) ◄—► (E)

The contents of registers H and L are exchanged with the contents of registers D and E.

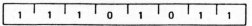

1	1	1	0	1	0	1	1

Cycles: 1
States: 4
Addressing: register
Flags: none

Arithmetic Group:

This group of instructions performs arithmetic operations on data in registers and memory.

Unless indicated otherwise, all instructions in this group affect the Zero, Sign, Parity, Carry, and Auxiliary Carry flags according to the standard rules.

All subtraction operations are performed via two's complement arithmetic and set the carry flag to one to indicate a borrow and clear it to indicate no borrow.

ADD r (Add Register)
$(A) \leftarrow (A) + (r)$

The content of register r is added to the content of the accumulator. The result is placed in the accumulator.

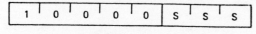

Cycles:	1	
States:	4	
Addressing:	register	
Flags:	Z,S,P,CY,AC	

ADD M (Add memory)
$(A) \leftarrow (A) + ((H) (L))$

The content of the memory location whose address is contained in the H and L registers is added to the content of the accumulator. The result is placed in the accumulator.

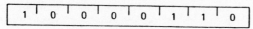

Cycles:	2	
States:	7	
Addressing:	reg. indirect	
Flags:	Z,S,P,CY,AC	

ADI data (Add immediate)
$(A) \leftarrow (A) + (byte 2)$

The content of the second byte of the instruction is added to the content of the accumulator. The result is placed in the accumulator.

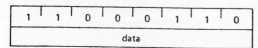

Cycles:	2	
States:	7	
Addressing:	immediate	
Flags:	Z,S,P,CY,AC	

ADC r (Add Register with carry)
$(A) \leftarrow (A) + (r) + (CY)$

The content of register r and the content of the carry bit are added to the content of the accumulator. The result is placed in the accumulator.

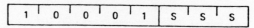

Cycles:	1	
States:	4	
Addressing:	register	
Flags:	Z,S,P,CY,AC	

ADC M (Add memory with carry)
$(A) \leftarrow (A) + ((H) (L)) + (CY)$

The content of the memory location whose address is contained in the H and L registers and the content of the CY flag are added to the accumulator. The result is placed in the accumulator.

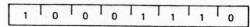

Cycles:	2	
States:	7	
Addressing:	reg. indirect	
Flags:	Z,S,P,CY,AC	

ACI data (Add immediate with carry)
$(A) \leftarrow (A) + (byte 2) + (CY)$

The content of the second byte of the instruction and the content of the CY flag are added to the contents of the accumulator. The result is placed in the accumulator.

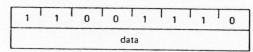

Cycles:	2	
States:	7	
Addressing:	immediate	
Flags:	Z,S,P,CY,AC	

SUB r (Subtract Register)
$(A) \leftarrow (A) - (r)$

The content of register r is subtracted from the content of the accumulator. The result is placed in the accumulator.

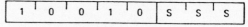

Cycles:	1	
States:	4	
Addressing:	register	
Flags:	Z,S,P,CY,AC	

SUB M   (Subtract memory)

(A) ◄— (A) − ((H) (L))

The content of the memory location whose address is contained in the H and L registers is subtracted from the content of the accumulator. The result is placed in the accumulator.

1	0	0	1	0	1	1	0

Cycles: 2
States: 7
Addressing: reg. indirect
Flags: Z,S,P,CY,AC

SUI data   (Subtract immediate)

(A) ◄— (A) − (byte 2)

The content of the second byte of the instruction is subtracted from the content of the accumulator. The result is placed in the accumulator.

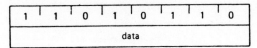

1	1	0	1	0	1	1	0
data							

Cycles: 2
States: 7
Addressing: immediate
Flags: Z,S,P,CY,AC

SBB r   (Subtract Register with borrow)

(A) ◄— (A) − (r) − (CY)

The content of register r and the content of the CY flag are both subtracted from the accumulator. The result is placed in the accumulator.

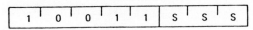

1	0	0	1	1	S	S	S

Cycles: 1
States: 4
Addressing: register
Flags: Z,S,P,CY,AC

SBB M   (Subtract memory with borrow)

(A) ◄— (A) − ((H) (L)) − (CY)

The content of the memory location whose address is contained in the H and L registers and the content of the CY flag are both subtracted from the accumulator. The result is placed in the accumulator.

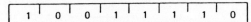

1	0	0	1	1	1	1	0

Cycles: 2
States: 7
Addressing: reg. indirect
Flags: Z,S,P,CY,AC

SBI data   (Subtract immediate with borrow)

(A) ◄— (A) − (byte 2) − (CY)

The contents of the second byte of the instruction and the contents of the CY flag are both subtracted from the accumulator. The result is placed in the accumulator.

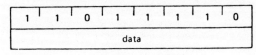

1	1	0	1	1	1	1	0
data							

Cycles: 2
States: 7
Addressing: immediate
Flags: Z,S,P,CY,AC

INR r   (Increment Register)

(r) ◄— (r) + 1

The content of register r is incremented by one. Note: All condition flags except CY are affected.

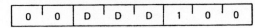

0	0	D	D	D	1	0	0

Cycles: 1
States: 4
Addressing: register
Flags: Z,S,P,AC

INR M   (Increment memory)

((H) (L)) ◄— ((H) (L)) + 1

The content of the memory location whose address is contained in the H and L registers is incremented by one. Note: All condition flags except CY are affected.

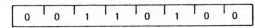

0	0	1	1	0	1	0	0

Cycles: 3
States: 10
Addressing: reg. indirect
Flags: Z,S,P,AC

DCR r   (Decrement Register)

(r) ◄— (r) − 1

The content of register r is decremented by one. Note: All condition flags except CY are affected.

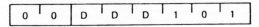

0	0	D	D	D	1	0	1

Cycles: 1
States: 4
Addressing: register
Flags: Z,S,P,AC

DCR M (Decrement memory)

((H) (L)) ◄── ((H) (L)) − 1

The content of the memory location whose address is contained in the H and L registers is decremented by one. Note: All condition flags **except CY** are affected.

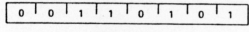

| 0 | 0 | 1 | 1 | 0 | 1 | 0 | 1 |

Cycles: 3
States: 10
Addressing: reg. indirect
Flags: Z,S,P,AC

INX rp (Increment register pair)

(rh) (rl) ◄── (rh) (rl) + 1

The content of the register pair rp is incremented by one. Note: **No condition flags are affected.**

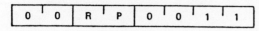

| 0 | 0 | R | P | 0 | 0 | 1 | 1 |

Cycles: 1
States: 6
Addressing: register
Flags: none

DCX rp (Decrement register pair)

(rh) (rl) ◄── (rh) (rl) − 1

The content of the register pair rp is decremented by one. Note: **No condition flags are affected.**

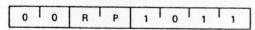

| 0 | 0 | R | P | 1 | 0 | 1 | 1 |

Cycles: 1
States: 6
Addressing: register
Flags: none

DAD rp (Add register pair to H and L)

(H) (L) ◄── (H) (L) + (rh) (rl)

The content of the register pair rp is added to the content of the register pair H and L. The result is placed in the register pair H and L. Note: Only the **CY** flag is affected. It is set if there is a carry out of the double precision add; otherwise it is reset.

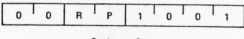

| 0 | 0 | R | P | 1 | 0 | 0 | 1 |

Cycles: 3
States: 10
Addressing: register
Flags: CY

DAA (Decimal Adjust Accumulator)

The eight-bit number in the accumulator is adjusted to form two four-bit Binary-Coded-Decimal digits by the following process:

1. If the value of the least significant 4 bits of the accumulator is greater than 9 or if the AC flag is set, 6 is added to the accumulator.

2. If the value of the most significant 4 bits of the accumulator is now greater than 9, or if the CY flag is set, 6 is added to the most significant 4 bits of the accumulator.

NOTE: All flags are affected.

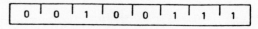

| 0 | 0 | 1 | 0 | 0 | 1 | 1 | 1 |

Cycles: 1
States: 4
Flags: Z,S,P,CY,AC

Logical Group:

This group of instructions performs logical (Boolean) operations on data in registers and memory and on condition flags.

Unless indicated otherwise, all instructions in this group affect the Zero, Sign, Parity, Auxiliary Carry, and Carry flags according to the standard rules.

ANA r (AND Register)

(A) ◄── (A) ∧ (r)

The content of register r is logically anded with the content of the accumulator. The result is placed in the accumulator. **The CY flag is cleared and AC is set.**

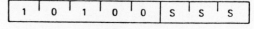

| 1 | 0 | 1 | 0 | 0 | S | S | S |

Cycles: 1
States: 4
Addressing: register
Flags: Z,S,P,CY,AC

ANA M (AND memory)

(A) ◄── (A) ∧ ((H) (L))

The contents of the memory location whose address is contained in the H and L registers is logically anded with the content of the accumulator. The result is placed in the accumulator. **The CY flag is cleared and AC is set.**

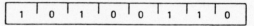

| 1 | 0 | 1 | 0 | 0 | 1 | 1 | 0 |

Cycles: 2
States: 7
Addressing: reg. indirect
Flags: Z,S,P,CY,AC

ANI data   (AND immediate)

(A) ◄── (A) ∧ (byte 2)

The content of the second byte of the instruction is logically anded with the contents of the accumulator. The result is placed in the accumulator. **The CY flag is cleared and AC is set.**

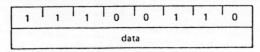

Cycles:	2
States:	7
Addressing:	immediate
Flags:	Z,S,P,CY,AC

XRA r   (Exclusive OR Register)

(A) ◄── (A) ∀ (r)

The content of register r is exclusive-or'd with the content of the accumulator. The result is placed in the accumulator. **The CY and AC flags are cleared.**

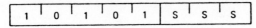

Cycles:	1
States:	4
Addressing:	register
Flags:	Z,S,P,CY,AC

XRA M   (Exclusive OR Memory)

(A) ◄── (A) ∀ ((H) (L))

The content of the memory location whose address is contained in the H and L registers is exclusive-OR'd with the content of the accumulator. The result is placed in the accumulator. **The CY and AC flags are cleared.**

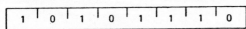

Cycles:	2
States:	7
Addressing:	reg. indirect
Flags:	Z,S,P,CY,AC

XRI data   (Exclusive OR immediate)

(A) ◄── (A) ∀ (byte 2)

The content of the second byte of the instruction is exclusive-OR'd with the content of the accumulator. The result is placed in the accumulator. **The CY and AC flags are cleared.**

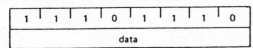

Cycles:	2
States:	7
Addressing:	immediate
Flags:	Z,S,P,CY,AC

ORA r   (OR Register)

(A) ◄── (A) V (r)

The content of register r is inclusive-OR'd with the content of the accumulator. The result is placed in the accumulator. **The CY and AC flags are cleared.**

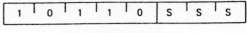

Cycles:	1
States:	4
Addressing:	register
Flags:	Z,S,P,CY,AC

ORA M   (OR memory)

(A) ◄── (A) V ((H) (L))

The content of the memory location whose address is contained in the H and L registers is inclusive-OR'd with the content of the accumulator. The result is placed in the accumulator. **The CY and AC flags are cleared.**

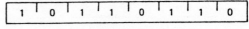

Cycles:	2
States:	7
Addressing:	reg. indirect
Flags:	Z,S,P,CY,AC

ORI data   (OR Immediate)

(A) ◄── (A) V (byte 2)

The content of the second byte of the instruction is inclusive-OR'd with the content of the accumulator. The result is placed in the accumulator. **The CY and AC flags are cleared.**

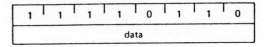

Cycles:	2
States:	7
Addressing:	immediate
Flags:	Z,S,P,CY,AC

CMP r   (Compare Register)

(A) − (r)

The content of register r is subtracted from the accumulator. The accumulator remains unchanged. The condition flags are set as a result of the subtraction. The Z flag is set to 1 if (A) = (r). The CY flag is set to 1 if (A) < (r).

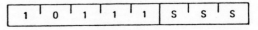

Cycles:	1
States:	4
Addressing:	register
Flags:	Z,S,P,CY,AC

CMP M   (Compare memory)

(A) − ((H) (L))

The content of the memory location whose address is contained in the H and L registers is subtracted from the accumulator. The accumulator remains unchanged. The condition flags are set as a result of the subtraction. The Z flag is set to 1 if (A) = ((H) (L)). The CY flag is set to 1 if (A) < ((H) (L)).

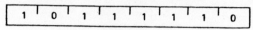

$$\begin{array}{|c|c|c|c|c|c|c|c|} \hline 1 & 0 & 1 & 1 & 1 & 1 & 1 & 0 \\ \hline \end{array}$$

Cycles:	2
States:	7
Addressing:	reg. indirect.
Flags:	Z,S,P,CY,AC

CPI data   (Compare immediate)

(A) − (byte 2)

The content of the second byte of the instruction is subtracted from the accumulator. The condition flags are set by the result of the subtraction. The Z flag is set to 1 if (A) = (byte 2). The CY flag is set to 1 if (A) < (byte 2).

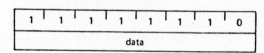

Cycles:	2
States:	7
Addressing:	immediate
Flags:	Z,S,P,CY,AC

RLC   (Rotate left)

$(A_{n+1}) \leftarrow (A_n)$; $(A_0) \leftarrow (A_7)$

$(CY) \leftarrow (A_7)$

The content of the accumulator is rotated left one position. The low order bit and the CY flag are both set to the value shifted out of the high order bit position. Only the CY flag is affected.

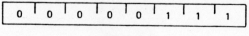

Cycles:	1
States:	4
Flags:	CY

RRC   (Rotate right)

$(A_n) \leftarrow (A_{n-1})$; $(A_7) \leftarrow (A_0)$

$(CY) \leftarrow (A_0)$

The content of the accumulator is rotated right one position. The high order bit and the CY flag are both set to the value shifted out of the low order bit position. Only the CY flag is affected.

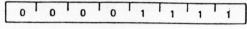

Cycles:	1
States:	4
Flags:	CY

RAL   (Rotate left through carry)

$(A_{n+1}) \leftarrow (A_n)$; $(CY) \leftarrow (A_7)$

$(A_0) \leftarrow (CY)$

The content of the accumulator is rotated left one position through the CY flag. The low order bit is set equal to the CY flag and the CY flag is set to the value shifted out of the high order bit. Only the CY flag is affected.

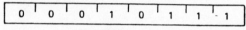

Cycles:	1
States:	4
Flags:	CY

RAR   (Rotate right through carry)

$(A_n) \leftarrow (A_{n+1})$; $(CY) \leftarrow (A_0)$

$(A_7) \leftarrow (CY)$

The content of the accumulator is rotated right one position through the CY flag. The high order bit is set to the CY flag and the CY flag is set to the value shifted out of the low order bit. Only the CY flag is affected.

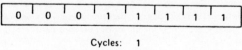

Cycles:	1
States:	4
Flags:	CY

CMA   (Complement accumulator)

(A) ← $\overline{(A)}$

The contents of the accumulator are complemented (zero bits become 1, one bits become 0). No flags are affected.

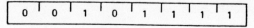

Cycles:	1
States:	4
Flags:	none

CMC (Complement carry)

(CY) ◄— (C̄Ȳ)

The CY flag is complemented. No other flags are affected.

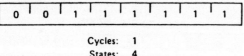

0	0	1	1	1	1	1	1

Cycles: 1
States: 4
Flags: CY

STC (Set carry)

(CY) ◄— 1

The CY flag is set to 1. No other flags are affected.

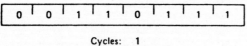

0	0	1	1	0	1	1	1

Cycles: 1
States: 4
Flags: CY

Branch Group:

This group of instructions alter normal sequential program flow.

Condition flags are not affected by any instruction in this group.

The two types of branch instructions are unconditional and conditional. Unconditional transfers simply perform the specified operation on register PC (the program counter). Conditional transfers examine the status of one of the four processor flags to determine if the specified branch is to be executed. The conditions that may be specified are as follows:

CONDITION		CCC
NZ	— not zero (Z = 0)	000
Z	— zero (Z = 1)	001
NC	— no carry (CY = 0)	010
C	— carry (CY = 1)	011
PO	— parity odd (P = 0)	100
PE	— parity even (P = 1)	101
P	— plus (S = 0)	110
M	— minus (S = 1)	111

JMP addr (Jump)

(PC) ◄— (byte 3) (byte 2)

Control is transferred to the instruction whose address is specified in byte 3 and byte 2 of the current instruction.

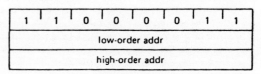

1	1	0	0	0	0	1	1
low-order addr							
high-order addr							

Cycles: 3
States: 10
Addressing: immediate
Flags: none

Jcondition addr (Conditional jump)

If (CCC),

(PC) ◄— (byte 3) (byte 2)

If the specified condition is true, control is transferred to the instruction whose address is specified in byte 3 and byte 2 of the current instruction; otherwise, control continues sequentially.

1	1	C	C	C	0	1	0
low-order addr							
high-order addr							

Cycles: 2/3
States: 7/10
Addressing: immediate
Flags: none

CALL addr (Call)

((SP) − 1) ◄— (PCH)

((SP) − 2) ◄— (PCL)

(SP) ◄— (SP) − 2

(PC) ◄— (byte 3) (byte 2)

The high-order eight bits of the next instruction address are moved to the memory location whose address is one less than the content of register SP. The low-order eight bits of the next instruction address are moved to the memory location whose address is two less than the content of register SP. The content of register SP is decremented by 2. Control is transferred to the instruction whose address is specified in byte 3 and byte 2 of the current instruction.

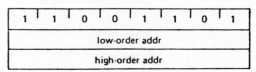

1	1	0	0	1	1	0	1
low-order addr							
high-order addr							

Cycles: 5
States: 18
Addressing: immediate/reg. indirect
Flags: none

Ccondition addr   (Condition call)

If (CCC),

((SP) − 1) ← (PCH)

((SP) − 2) ← (PCL)

(SP) ← (SP) − 2

(PC) ← (byte 3) (byte 2)

If the specified condition is true, the actions specified in the CALL instruction (see above) are performed; otherwise, control continues sequentially.

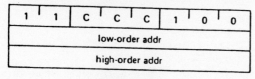

1	1	C	C	C	1	0	0
low-order addr								
high-order addr								

Cycles: 2/5

States: 9/18

Addressing: immediate/reg. indirect

Flags: none

RET   (Return)

(PCL) ← ((SP));

(PCH) ← ((SP) + 1);

(SP) ← (SP) + 2;

The content of the memory location whose address is specified in register SP is moved to the low-order eight bits of register PC. The content of the memory location whose address is one more than the content of register SP is moved to the high-order eight bits of register PC. The content of register SP is incremented by 2.

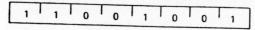

1	1	0	0	1	0	0	1

Cycles: 3

States: 10

Addressing: reg. indirect

Flags: none

Rcondition   (Conditional return)

If (CCC),

(PCL) ← ((SP))

(PCH) ← ((SP) + 1)

(SP) ← (SP) + 2

If the specified condition is true, the actions specified in the RET instruction (see above) are performed; otherwise, control continues sequentially.

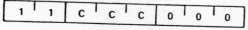

1	1	C	C	C	0	0	0

Cycles: 1/3

States: 6/12

Addressing: reg. indirect

Flags: none

RST n   (Restart)

((SP) − 1) ← (PCH)

((SP) − 2) ← (PCL)

(SP) ← (SP) − 2

(PC) ← 8 • (NNN)

The high-order eight bits of the next instruction address are moved to the memory location whose address is one less than the content of register SP. The low-order eight bits of the next instruction address are moved to the memory location whose address is two less than the content of register SP. The content of register SP is decremented by two. Control is transferred to the instruction whose address is eight times the content of NNN.

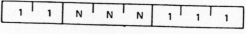

1	1	N	N	N	1	1	1

Cycles: 3

States: 12

Addressing: reg. indirect

Flags: none

15	14	13	12	11	10	9	8	7	6	5	4	3	2	1	0
0	0	0	0	0	0	0	0	0	0	N	N	N	0	0	0

Program Counter After Restart

PCHL   (Jump H and L indirect − move H and L to PC)

(PCH) ← (H)

(PCL) ← (L)

The content of register H is moved to the high-order eight bits of register PC. The content of register L is moved to the low-order eight bits of register PC.

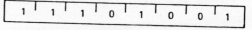

1	1	1	0	1	0	0	1

Cycles: 1

States: 6

Addressing: register

Flags: none

Stack, I/O, and Machine Control Group:

This group of instructions performs I/O, manipulates the Stack, and alters internal control flags.

Unless otherwise specified, condition flags are not affected by any instructions in this group.

FLAG WORD

D_7	D_6	D_5	D_4	D_3	D_2	D_1	D_0
S	Z	X	AC	X	P	X	CY

X: undefined

PUSH rp   (Push)

$((SP) - 1) \leftarrow (rh)$

$((SP) - 2) \leftarrow (rl)$

$(SP) \leftarrow (SP) - 2$

The content of the high-order register of register pair rp is moved to the memory location whose address is one less than the content of register SP. The content of the low-order register of register pair rp is moved to the memory location whose address is two less than the content of register SP. The content of register SP is decremented by 2. Note: Register pair rp = SP may not be specified.

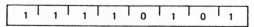

1	1	R	P	0	1	0	1

Cycles: **3**
States: **12**
Addressing: reg. indirect
Flags: none

PUSH PSW   (Push processor status word)

$((SP) - 1) \leftarrow (A)$

$((SP) - 2)_0 \leftarrow (CY) , ((SP) - 2)_1 \leftarrow 1$

$((SP) - 2)_2 \leftarrow (P) , \quad ((SP) - 2)_3 \leftarrow 0$

$((SP) - 2)_4 \leftarrow (AC) , ((SP) - 2)_5 \leftarrow 0$

$((SP) - 2)_6 \leftarrow (Z) , \quad ((SP) - 2)_7 \leftarrow (S)$

$(SP) \leftarrow (SP) - 2$

The content of register A is moved to the memory location whose address is one less than register SP. The contents of the condition flags are assembled into a processor status word and the word is moved to the memory location whose address is two less than the content of register SP. The content of register SP is decremented by two.

1	1	1	1	0	1	0	1

Cycles: **3**
States: **12**
Addressing: reg. indirect
Flags: none

POP rp   (Pop)

$(rl) \leftarrow ((SP))$

$(rh) \leftarrow ((SP) + 1)$

$(SP) \leftarrow (SP) + 2$

The content of the memory location, whose address is specified by the content of register SP, is moved to the low-order register of register pair rp. The content of the memory location, whose address is one more than the content of register SP, is moved to the high-order register of register pair rp. The content of register SP is incremented by 2. Note: Register pair rp = SP may not be specified.

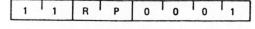

1	1	R	P	0	0	0	1

Cycles: **3**
States: **10**
Addressing: reg. indirect
Flags: none

POP PSW   (Pop processor status word)

$(CY) \leftarrow ((SP))_0$

$(P) \leftarrow ((SP))_2$

$(AC) \leftarrow ((SP))_4$

$(Z) \leftarrow ((SP))_6$

$(S) \leftarrow ((SP))_7$

$(A) \leftarrow ((SP) + 1)$

$(SP) \leftarrow (SP) + 2$

The content of the memory location whose address is specified by the content of register SP is used to restore the condition flags. The content of the memory location whose address is one more than the content of register SP is moved to register A. The content of register SP is incremented by 2.

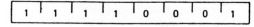

1	1	1	1	0	0	0	1

Cycles: **3**
States: **10**
Addressing: reg. indirect
Flags: Z,S,P,CY,AC

XTHL (Exchange stack top with H and L)

(L) ⟷ ((SP))

(H) ⟷ ((SP) + 1)

The content of the L register is exchanged with the content of the memory location whose address is specified by the content of register SP. The content of the H register is exchanged with the content of the memory location whose address is one more than the content of register SP.

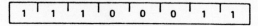

Cycles: 5
States: 16
Addressing: reg. indirect
Flags: none

SPHL (Move HL to SP)

(SP) ⟵ (H) (L)

The contents of registers H and L (16 bits) are moved to register SP.

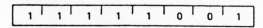

Cycles: 1
States: 6
Addressing: register
Flags: none

IN port   (Input)

(A) ⟵ (data)

The data placed on the eight bit bi-directional data bus by the specified port is moved to register A.

Cycles: 3
States: 10
Addressing: direct
Flags: none

OUT port   (Output)

(data) ⟵ (A)

The content of register A is placed on the eight bit bi-directional data bus for transmission to the specified port.

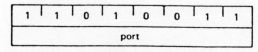

Cycles: 3
States: 10
Addressing: direct
Flags: none

EI   (Enable interrupts)

The interrupt system is enabled following the execution of the next instruction.

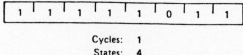

Cycles: 1
States: 4
Flags: none

DI   (Disable interrupts)

The interrupt system is disabled immediately following the execution of the DI instruction.

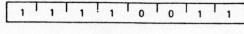

Cycles: 1
States: 4
Flags: none

HLT   (Halt)

The processor is stopped. The registers and flags are unaffected.

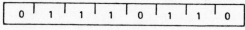

Cycles: 1
States: 5
Flags: none

NOP   (No op)

No operation is performed. The registers and flags are unaffected.

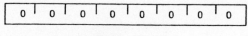

Cycles: 1
States: 4
Flags: none

RIM (Read Interrupt Mask)

The accumulator is loaded with the restart interrupt masks, any pending interrupts, and the contents of the serial input data line (SID).

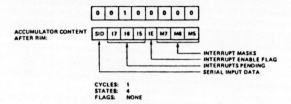

SIM (Set Interrupt Masks)

The contents of the accumulator will be used in programming the restart interrupt masks. Bits 0–2 will set/reset the mask bit for RST 5.5, 6.5, 7.5 of the interrupt mask register, if bit 3 is 1 ("set"). Bit 3 is a "Mask Set Enable" control.

Setting the mask (i.e. masked bit = 1) disables the corresponding interrupt.

	Set	Reset
RST 5.5 MASK	if bit 0 = 1	if bit 0 = 0
RST 6.5 MASK	bit 1 = 1	bit 1 = 0
RST 7.5 MASK	bit 2 = 1	bit 2 = 0

RST 7.5, whether masked or not, will be reset if bit 4 = 1.

RESET IN input (pin 36) will set all RST MASKs, and reset/disable all interrupts.

SIM can, also, load the SOD output latch. Accumulator bit 7 is loaded into the SOD latch if bit 6 is set. The latch is unaffected if bit 6 is a zero. RESET IN input sets the SOD latch to zero.

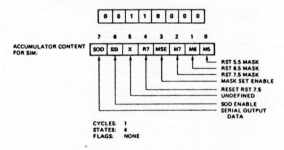

8085 INSTRUCTION SET

Summary of Processor Instructions

Mnemonic	Description	D_7	D_6	D_5	D_4	D_3	D_2	D_1	D_0
MOV r1, r2	Move register to register	0	1	D	D	D	S	S	S
MOV M, r	Move register to memory	0	1	1	1	0	S	S	S
MOV r, M	Move memory to register	0	1	D	D	D	1	1	0
HLT	Halt	0	1	1	1	0	1	1	0
MVI r	Move immediate register	0	0	D	D	D	1	1	0
MVI M	Move immediate memory	0	0	1	1	0	1	1	0
INR r	Increment register	0	0	D	D	D	1	0	0
DCR r	Decrement register	0	0	D	D	D	1	0	1
INR M	Increment memory	0	0	1	1	0	1	0	0
DCR M	Decrement memory	0	0	1	1	0	1	0	1
ADD r	Add register to A	1	0	0	0	0	S	S	S
ADC r	Add register to A with carry	1	0	0	0	1	S	S	S
SUB r	Subtract register from A	1	0	0	1	0	S	S	S
SBB r	Subtract register from A with borrow	1	0	0	1	1	S	S	S
ANA r	And register with A	1	0	1	0	0	S	S	S
XRA r	Exclusive Or register with A	1	0	1	0	1	S	S	S
ORA r	Or register with A	1	0	1	1	0	S	S	S
CMP r	Compare register with A	1	0	1	1	1	S	S	S
ADD M	Add memory to A	1	0	0	0	0	1	1	0
ADC M	Add memory to A with carry	1	0	0	0	1	1	1	0
SUB M	Subtract memory from A	1	0	0	1	0	1	1	0
SBB M	Subtract memory from A with borrow	1	0	0	1	1	1	1	0
ANA M	And memory with A	1	0	1	0	0	1	1	0
XRA M	Exclusive Or memory with A	1	0	1	0	1	1	1	0
ORA M	Or memory with A	1	0	1	1	0	1	1	0
CMP M	Compare memory with A	1	0	1	1	1	1	1	0
ADI	Add immediate to A	1	1	0	0	0	1	1	0
ACI	Add immediate to A with carry	1	1	0	0	1	1	1	0
SUI	Subtract immediate from A	1	1	0	1	0	1	1	0
SBI	Subtract immediate from A with borrow	1	1	0	1	1	1	1	0
ANI	And immediate with A	1	1	1	0	0	1	1	0
XRI	Exclusive Or immediate with A	1	1	1	0	1	1	1	0
ORI	Or immediate with A	1	1	1	1	0	1	1	0
CPI	Compare immediate with A	1	1	1	1	1	1	1	0
RLC	Rotate A left	0	0	0	0	0	1	1	1
RRC	Rotate A right	0	0	0	0	1	1	1	1
RAL	Rotate A left through carry	0	0	0	1	0	1	1	1
RAR	Rotate A right through carry	0	0	0	1	1	1	1	1
JMP	Jump unconditional	1	1	0	0	0	0	1	1
JC	Jump on carry	1	1	0	1	1	0	1	0
JNC	Jump on no carry	1	1	0	1	0	0	1	0
JZ	Jump on zero	1	1	0	0	1	0	1	0
JNZ	Jump on no zero	1	1	0	0	0	0	1	0
JP	Jump on positive	1	1	1	1	0	0	1	0
JM	Jump on minus	1	1	1	1	1	0	1	0
JPE	Jump on parity even	1	1	1	0	1	0	1	0
JPO	Jump on parity odd	1	1	1	0	0	0	1	0
CALL	Call unconditional	1	1	0	0	1	1	0	1
CC	Call on carry	1	1	0	1	1	1	0	0
CNC	Call on no carry	1	1	0	1	0	1	0	0
CZ	Call on zero	1	1	0	0	1	1	0	0
CNZ	Call on no zero	1	1	0	0	0	1	0	0
CP	Call on positive	1	1	1	1	0	1	0	0
CM	Call on minus	1	1	1	1	1	1	0	0
CPE	Call on parity even	1	1	1	0	1	1	0	0
CPO	Call on parity odd	1	1	1	0	0	1	0	0
RET	Return	1	1	0	0	1	0	0	1
RC	Return on carry	1	1	0	1	1	0	0	0
RNC	Return on no carry	1	1	0	1	0	0	0	0

Mnemonic	Description	D_7	D_6	D_5	D_4	D_3	D_2	D_1	D_0
RZ	Return on zero	1	1	0	0	1	0	0	0
RNZ	Return on no zero	1	1	0	0	0	0	0	0
RP	Return on positive	1	1	1	1	0	0	0	0
RM	Return on minus	1	1	1	1	1	0	0	0
RPE	Return on parity even	1	1	1	0	1	0	0	0
RPO	Return on parity odd	1	1	1	0	0	0	0	0
RST	Restart	1	1	A	A	A	1	1	1
IN	Input	1	1	0	1	1	0	1	1
OUT	Output	1	1	0	1	0	0	1	1
LXI B	Load immediate register Pair B & C	0	0	0	0	0	0	0	1
LXI D	Load immediate register Pair D & E	0	0	0	1	0	0	0	1
LXI H	Load immediate register Pair H & L	0	0	1	0	0	0	0	1
LXI SP	Load immediate stack pointer	0	0	1	1	0	0	0	1
PUSH B	Push register Pair B & C on stack	1	1	0	0	0	1	0	1
PUSH D	Push register Pair D & E on stack	1	1	0	1	0	1	0	1
PUSH H	Push register Pair H & L on stack	1	1	1	0	0	1	0	1
PUSH PSW	Push A and Flags on stack	1	1	1	1	0	1	0	1
POP B	Pop register pair B & C off stack	1	1	0	0	0	0	0	1
POP D	Pop register pair D & E off stack	1	1	0	1	0	0	0	1
POP H	Pop register pair H & L off stack	1	1	1	0	0	0	0	1
POP PSW	Pop A and Flags off stack	1	1	1	1	0	0	0	1
STA	Store A direct	0	0	1	1	0	0	1	0
LDA	Load A direct	0	0	1	1	1	0	1	0
XCHG	Exchange D & E, H & L Registers	1	1	1	0	1	0	1	1
XTHL	Exchange top of stack, H & L	1	1	1	0	0	0	1	1
SPHL	H & L to stack pointer	1	1	1	1	1	0	0	1
PCHL	H & L to program counter	1	1	1	0	1	0	0	1
DAD B	Add B & C to H & L	0	0	0	0	1	0	0	1
DAD D	Add D & E to H & L	0	0	0	1	1	0	0	1
DAD H	Add H & L to H & L	0	0	1	0	1	0	0	1
DAD SP	Add stack pointer to H & L	0	0	1	1	1	0	0	1
STAX B	Store A indirect	0	0	0	0	0	0	1	0
STAX D	Store A indirect	0	0	0	1	0	0	1	0
LDAX B	Load A indirect	0	0	0	0	1	0	1	0
LDAX D	Load A indirect	0	0	0	1	1	0	1	0
INX B	Increment B & C registers	0	0	0	0	0	0	1	1
INX D	Increment D & E registers	0	0	0	1	0	0	1	1
INX H	Increment H & L registers	0	0	1	0	0	0	1	1
INX SP	Increment stack pointer	0	0	1	1	0	0	1	1
DCX B	Decrement B & C	0	0	0	0	1	0	1	1
DCX D	Decrement D & E	0	0	0	1	1	0	1	1
DCX H	Decrement H & L	0	0	1	0	1	0	1	1
DCX SP	Decrement stack pointer	0	0	1	1	1	0	1	1
CMA	Complement A	0	0	1	0	1	1	1	1
STC	Set carry	0	0	1	1	0	1	1	1
CMC	Complement carry	0	0	1	1	1	1	1	1
DAA	Decimal adjust A	0	0	1	0	0	1	1	1
SHLD	Store H & L direct	0	0	1	0	0	0	1	0
LHLD	Load H & L direct	0	0	1	0	1	0	1	0
EI	Enable Interrupts	1	1	1	1	1	0	1	1
DI	Disable interrupt	1	1	1	1	0	0	1	1
NOP	No-operation	0	0	0	0	0	0	0	0
RIM	Read Interrupt Mask	0	0	1	0	0	0	0	0
SIM	Set Interrupt Mask	0	0	1	1	0	0	0	0

NOTE   DDD or SSS – 000 B – 001 C – 010 D – 011 E – 100 H – 101 L – 110 Memory – 111 A.

Appendix 2 Data on the Zilog Z80 microprocessor

The following pages are reproduced, by kind permission of Zilog Ltd, direct from the Zilog Data Book on the Z80 microprocessor. They give an overview, hardware description and instruction set for the Z80 microprocessor.

2.0 Z-80 CPU ARCHITECTURE

A block diagram of the internal architecture of the Z-80 CPU is shown in figure 2.0-1. The diagram shows all of the major elements in the CPU and it should be referred to throughout the following description.

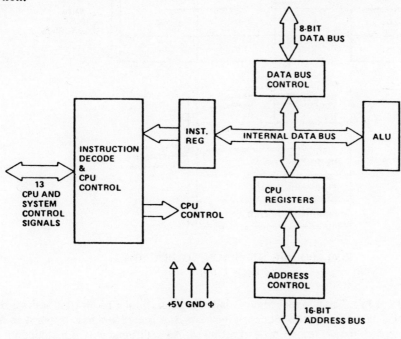

Z-80 CPU BLOCK DIAGRAM
FIGURE 2.0-1

2.1 CPU REGISTERS

The Z-80 CPU contains 208 bits of R/W memory that are accessible to the programmer. Figure 2.0-2 illustrates how this memory is configured into eighteen 8-bit registers and four 16-bit registers. All Z-80 registers are implemented using static RAM. The registers include two sets of six general purpose registers that may be used individually as 8-bit registers or in pairs as 16-bit registers. There are also two sets of accumulator and flag registers.

Special Purpose Registers

1. **Program Counter (PC).** The program counter holds the 16-bit address of the current instruction being fetched from memory. The PC is automatically incremented after its contents have been transferred to the address lines. When a program jump occurs the new value is automatically placed in the PC, overriding the incrementer.

2. **Stack Pointer (SP).** The stack pointer holds the 16-bit address of the current top of a stack located anywhere in external system RAM memory. The external stack memory is organized as a last-in first-out (LIFO) file. Data can be pushed onto the stack from specific CPU registers or popped off of the stack into specific CPU registers through the execution of PUSH and POP instructions. The data popped from the stack is always the last data pushed onto it. The stack allows simple implementation of multiple level interrupts, unlimited subroutine nesting and simplification of many types of data manipulation.

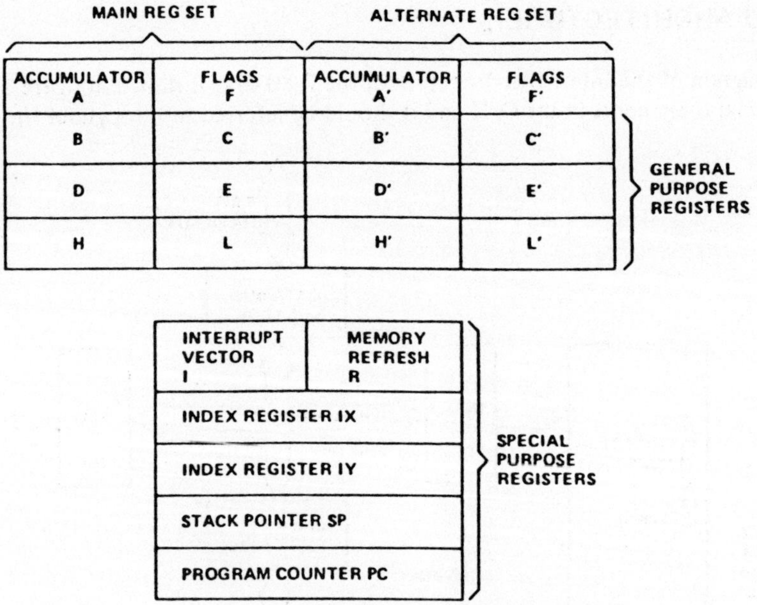

MAIN REG SET   ALTERNATE REG SET

ACCUMULATOR A	FLAGS F	ACCUMULATOR A'	FLAGS F'
B	C	B'	C'
D	E	D'	E'
H	L	H'	L'

GENERAL PURPOSE REGISTERS

INTERRUPT VECTOR I	MEMORY REFRESH R
INDEX REGISTER IX	
INDEX REGISTER IY	
STACK POINTER SP	
PROGRAM COUNTER PC	

SPECIAL PURPOSE REGISTERS

Z-80 CPU REGISTER CONFIGURATION
FIGURE 2.0-2

3. **Two Index Registers (IX & IY).** The two independent index registers hold a 16-bit base address that is used in indexed addressing modes. In this mode, an index register is used as a base to point to a region in memory from which data is to be stored or retrieved. An additional byte is included in indexed instructions to specify a displacement from this base. This displacement is specified as a two's complement signed integer. This mode of addressing greatly simplifies many types of programs, especially where tables of data are used.

4. **Interrupt Page Address Register (I).** The Z-80 CPU can be operated in a mode where an indirect call to any memory location can be achieved in response to an interrupt. The I Register is used for this purpose to store the high order 8-bits of the indirect address while the interrupting device provides the lower 8-bits of the address. This feature allows interrupt routines to be dynamically located anywhere in memory with absolute minimal access time to the routine.

5. **Memory Refresh Register (R).** The Z-80 CPU contains a memory refresh counter to enable dynamic memories to be used with the same ease as static memories. This 7-bit register is automatically incremented after each instruction fetch. The data in the refresh counter is sent out on the lower portion of the address bus along with a refresh control signal while the CPU is decoding and executing the fetched instruction. This mode of refresh is totally transparent to the programmer and does not slow down the CPU operation. The programmer can load the R register for testing purposes, but this register is normally not used by the programmer.

Accumulator and Flag Registers

The CPU includes two independent 8-bit accumulators and associated 8-bit flag registers. The accumulator holds the results of 8-bit arithmetic or logical operations while the flag register indicates specific conditions for 8 or 16-bit operations, such as indicating whether or not the result of an operation is equal to zero. The programmer selects the accumulator and flag pair that he wishes to work with with a single exchange instruction so that he may easily work with either pair.

3.0 Z-80 CPU PIN DESCRIPTION

The Z-80 CPU is packaged in an industry standard 40 pin Dual In-Line Package. The I/O pins are shown in figure 3.0-1 and the function of each is described below.

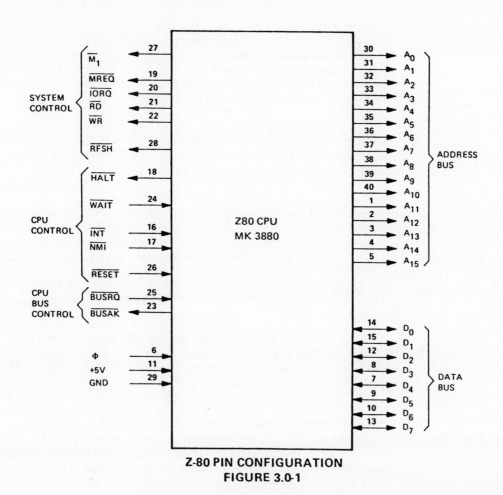

Z-80 PIN CONFIGURATION
FIGURE 3.0-1

A_0-A_{15}
(Address Bus)

Tri-state output, active high. A_0-A_{15} constitute a 16-bit address bus. The address bus provides the address for memory (up to 64K bytes) data exchanges and for I/O device data exchanges. I/O addressing uses the 8 lower address bits to allow the user to directly select up to 256 input or 256 output ports. A_0 is the least significant address bit. During refresh time, the lower 7 bits contain a valid refresh address.

D_0-D_7
(Data Bus)

Tri-state input/output, active high. D_0-D_7 constitute an 8-bit bidirectional data bus. The data bus is used for data exchanges with memory and I/O devices.

$\overline{M}_1$
(Machine Cycle one)

Output, active low. $\overline{M}_1$ indicates that the current machine cycle is the OP code fetch cycle of an instruction execution. Note that during execution of 2-byte op-codes, $\overline{M1}$ is generated as each op code byte is fetched. These two byte op-codes always begin with CBH, DDH, EDH or FDH. $\overline{M1}$ also occurs with $\overline{IORQ}$ to indicate an interrupt acknowledge cycle.

$\overline{MREQ}$
(Memory Request)

Tri-state output, active low. The memory request signal indicates that the address bus holds a valid address for a memory read or memory write operation.

$\overline{IORQ}$
(Input/Output Request)

Tri-state output, active low. The $\overline{IORQ}$ signal indicates that the lower half of the address bus holds a valid I/O address for a I/O read or write operation. An $\overline{IORQ}$ signal is also generated with an $\overline{M1}$ signal when an interrupt is being acknowledged to indicate that an interrupt response vector can be placed on the data bus. Interrupt Acknowledge operations occur during M_1 time while I/O operations never occur during M_1 time.

$\overline{RD}$
(Memory Read)

Tri-state output, active low. $\overline{RD}$ indicates that the CPU wants to read data from memory or an I/O device. The addressed I/O device or memory should use this signal to gate data onto the CPU data bus.

$\overline{WR}$
(Memory Write)

Tri-state output, active low. $\overline{WR}$ indicates that the CPU data bus holds valid data to be stored in the addressed memory or I/O device.

$\overline{RFSH}$
(Refresh)

Output, active low. $\overline{RFSH}$ indicates that the lower 7 bits of the address bus contain a refresh address for dynamic memories and the current $\overline{MREQ}$ signal should be used to do a refresh read to all dynamic memories.

$\overline{HALT}$
(Halt state)

Output, active low. $\overline{HALT}$ indicates that the CPU has executed a HALT software instruction and is awaiting either a non maskable or a maskable interrupt (with the mask enabled) before operation can resume. While halted, the CPU executes NOP's to maintain memory refresh activity.

$\overline{WAIT}$ *
(Wait)

Input, active low. $\overline{WAIT}$ indicates to the Z-80 CPU that the addressed memory or I/O devices are not ready for a data transfer. The CPU continues to enter wait states for as long as this signal is active. This signal allows memory or I/O devices of any speed to be synchronized to the CPU.

$\overline{INT}$
(Interrupt Request)

Input, active low. The Interrupt Request signal is generated by I/O devices. A request will be honored at the end of the current instruction if the internal software controlled interrupt enable flip-flop (IFF) is enabled and if the $\overline{BUSRQ}$ signal is not active. When the CPU accepts the interrupt, an acknowledge signal ($\overline{IORQ}$ during M_1 time) is sent out at the beginning of the next instruction cycle. The CPU can respond to an interrupt in three different modes that are described in detail in section 8.

---

*   While the Z80-CPU is in either a $\overline{WAIT}$ state or a Bus Acknowledge condition, Dynamic Memory Refresh will not occur.

$\overline{\text{NMI}}$
(Non Maskable
Interrupt)

Input, negative edge triggered. The non maskable interrupt request line has a higher priority than $\overline{\text{INT}}$ and is always recognized at the end of the current instruction, independent of the status of the interrupt enable flip-flop. $\overline{\text{NMI}}$ automatically forces the Z-80 CPU to restart to location 0066_H. The program counter is automatically saved in the external stack so that the user can return to the program that was interrupted. Note that continuous WAIT cycles can prevent the current instruction from ending, and that a $\overline{\text{BUSRQ}}$ will override a $\overline{\text{NMI}}$.

$\overline{\text{RESET}}$

Input, active low. $\overline{\text{RESET}}$ forces the program counter to zero and initializes the CPU. The CPU initialization includes:

1) Disable the interrupt enable flip-flop

2) Set Register I = 00_H

3) Set Register R = 00_H

4) Set Interrupt Mode 0

During reset time, the address bus and data bus go to a high impedance state and all control output signals go to the inactive state. No refresh occurs.

$\overline{\text{BUSRQ}}$
(Bus Request)

Input, active low. The bus request signal is used to request the CPU address bus, data bus and tri-state output control signals to go to a high impedance state so that other devices can control these buses. When $\overline{\text{BUSRQ}}$ is activated, the CPU will set these buses to a high impedance state as soon as the current CPU machine cycle is terminated.

$\overline{\text{BUSAK}}^*$
(Bus Acknowledge)

Output, active low. Bus acknowledge is used to indicate to the requesting device that the CPU address bus, data bus and tri-state control bus signals have been set to their high impedance state and the external device can now control these signals.

Φ

Single phase TTL level clock which requires only a 330 ohm pull-up resistor to +5 volts to meet all clock requirements.

5.0 Z-80 CPU INSTRUCTION SET

The Z-80 CPU can execute 158 different instruction types including all 78 of the 8080A CPU. The instructions can be broken down into the following major groups:

- Load and Exchange
- Block Transfer and Search
- Arithmetic and Logical
- Rotate and Shift
- Bit Manipulation (set, reset, test)

- Jump, Call and Return
- Input/Output
- Basic CPU Control

5.1   INTRODUCTION TO INSTRUCTION TYPES

The load instructions move data internally between CPU registers or between CPU registers and external memory. All of these instructions must specify a source location from which the data is to be moved and a destination location. The source location is not altered by a load instruction. Examples of load group instructions include moves between any of the general purpose registers such as move the data to Register B from Register C. This group also includes load immediate to any CPU register or to any external memory location. Other types of load instructions allow transfer between CPU registers and memory locations. The exchange instructions can trade the contents of two registers.

A unique set of block transfer instructions is provided in the Z-80. With a single instruction a block of memory of any size can be moved to any other location in memory. This set of block moves is extremely valuable when large strings of data must be processed. The Z-80 block search instructions are also valuable for this type of processing. With a single instruction, a block of external memory of any desired length can be searched for any 8-bit character. Once the character is found the instruction automatically terminates. Both the block transfer and the block search instructions can be interrupted during their execution so as to not occupy the CPU for long periods of time.

The arithmetic and logical instructions operate on data stored in the accumulator and other general purpose CPU registers or external memory locations. The results of the operations are placed in the accumulator and the appropriate flags are set according to the result of the operation. An example of an arithmetic operation is adding the accumulator to the contents of an external memory location. The results of the addition are placed in the accumulator. This group also includes 16-bit addition and subtraction between 16-bit CPU registers.

The bit manipulation instructions allow any bit in the accumulator, any general purpose register or any external memory location to be set, reset or tested with a single instruction. For example, the most significant bit of register H can be reset. This group is especially useful in control applications and for controlling software flags in general purpose programming.

The jump, call and return instructions are used to transfer between various locations in the user's program. This group uses several different techniques for obtaining the new program counter address from specific external memory locations. A unique type of jump is the restart instruction. This instruction actually contains the new address as a part of the 8-bit OP code. This is possible since only 8 separate addresses located in page zero of the external memory may be specified. Program jumps may also be achieved by loading register HL, IX or IY directly into the PC, thus allowing the jump address to be a complex function of the routine being executed.

Z80-CPU INSTRUCTION SET

ALPHABETICAL
ASSEMBLY MNEMONIC OPERATION

ADC HL,ss	Add with Carry Reg. pair ss to HL
ADC A,s	Add with carry operand s to Acc.
ADD A,n	Add value n to Acc.
ADD A,r	Add Reg. r to Acc.
ADD A,(HL)	Add location (HL) to Acc.
ADD A,(IX+d)	Add location (IX+d) to Acc.
ADD A,(IY+d)	Add location (IY+d) to Acc.
ADD HL,ss	Add Reg. pair ss to HL
ADD IX,pp	Add Reg. pair pp to IX
ADD IY,rr	Add Reg. pair rr to IY
AND s	Logical 'AND' of operand s and Acc.
BIT b,(HL)	Test BIT b of location (HL)
BIT b,(IX+d)	Test BIT b of location (IX+d)
BIT b,(IY+d)	Test BIT b of location (IY+d)
BIT b,r	Test BIT b of Reg. r
CALL cc,nn	Call subroutine at location nn if condition cc is true
CALL nn	Unconditional call subroutine at location nn
CCF	Complement carry flag
CP s	Compare operand s with Acc.
CPD	Compare location (HL) and Acc. decrement HL and BC
CPDR	Compare location (HL) and Acc. decrement HL and BC, repeat until BC=0
CPI	Compare location (HL) and Acc. increment HL and decrement BC
CPIR	Compare location (HL) and Acc. increment HL, decrement BC repeat until BC=0
CPL	Complement Acc. (1's comp)
DAA	Decimal adjust Acc.
DEC m	Decrement operand m
DEC IX	Decrement IX
DEC IY	Decrement IY
DEC ss	Decrement Reg. pair ss.
DI	Disable interrupts
DJNZ e	Decrement B and Jump relative if B≠0
EI	Enable interrupts
EX (SP),HL	Exchange the location (SP) and HL
EX (SP),IX	Exchange the location (SP) and IX
EX (SP),IY	Exchange the location (SP) and IY
EX AF,AF'	Exchange the contents of AF and AF'

EX DE,HL	Exchange the contents of DE and HL
EXX	Exchange the contents of BC,DE,HL with contents of BC',DE',HL' respectively
HALT	HALT (wait for interrupt or reset)
IM 0	Set interrupt mode 0
IM 1	Set interrupt mode 1
IM 2	Set interrupt mode 2
IN A,(n)	Load the Acc. with input from device n.
IN r,(C)	Load the Reg. r with input from device (C)
INC (HL)	Increment location (HL)
INC IX	Increment IX
INC (IX+d)	Increment location (IX+d
INC IY	Increment IY
INC (IY+d)	Increment location (IY+d)
INC r	Increment Reg. r
INC ss	Increment Reg. pair ss
IND	Load location (HL) with input from port (C), decrement HL and B
INDR	Load location (HL) with input from port (C), decrement HL and decrement B, repeat until B=0
INI	Load location (HL) with input from port (C); and increment HL and decrement B
INIR	Load location (HL) with input from port (C), increment HL and decrement B, repeat until B=0
JP (HL)	Unconditional Jump to (HL)
JP (IX)	Unconditional Jump to (IX)
JP (IY)	Unconditional Jump to (IY)
JP cc,nn	Jump to location nn if condition cc is true
JP nn	Unconditional jump to location nn
JR C,e	Jump relative to PC+e if carry=1
JR e	Unconditional Jump relative to PC+e
JR NC,e	Jump relative to PC+e if carry=0
JR NZ,e	Jump relative to PC+e if non zero (Z=0)
JR Z,e	Jump relative to PC+e if zero (Z=1)
LD A,(BC)	Load Acc. with location (BC)
LD A,(DE)	Load Acc. with location (DE)
LD A,I	Load Acc. with I
LD A,(nn)	Load Acc. with location nn
LD A,R	Load Acc. with Reg. R
LD (BC),A	Load location (BC) with Acc.

LD (DE),A	Load location (DE) with Acc.
LD (HL),n	Load location (HL) with value n
LD dd,nn	Load Reg. pair dd with value nn
LD dd,(nn)	Load Reg. pair dd with location (nn)
LD HL,(nn)	Load HL with location (nn)
LD (HL),r	Load location (HL) with Reg. r
LD I,A	Load I with Acc.
LF IX,nn	Load IX with value nn
LD IX,(nn)	Load IX with location (nn)
LD (IX+d),n	Load location (IX+d) with value n
LD (IX+d),r	Load location (IX+d) with Reg. r
LD IY,nn	Load IY with value nn
LD IY,(nn)	Load IY with location (nn)
LD (IY+d),n	Load location (IY+d) with value n
LD (IY+d),r	Load location (IY+d) with Reg. r
LD (nn),A	Load location (nn) with Acc.
LD (nn),dd	Load location (nn) with Reg. pair dd
LD (nn),HL	Load location (nn) with HL
LD (nn),IX	Load location (nn) with IX
LD (nn),IY	Load location (nn) with IY
LD R,A	Load R with Acc.
LD r,(HL)	Load Reg. r with location (HL)
LD r,(IX+d)	Load Reg. r with location (IX+d)
LD r,(IY+d)	Load Reg. r with location (IY+d)
LD r,n	Load Reg. r with value n
LD r,r´	Load Reg. r with Reg. r´
LD SP,HL	Load SP with HL
LD SP,IX	Load SP with IX
LD SP,IY	Load SP with IY
LDD	Load location (DE) with location (HL), decrement DE,HL and BC
LDDR	Load location (DE) with location (HL), decrement DE,HL and BC; repeat until BC=0
LDI	Load location (DE) with location (HL), increment DE,HL, decrement BC
LDIR	Load location (DE) with location (HL), increment DE,HL, decrement BC and repeat until BC=0
NEG	Negate Acc. (2´s complement)
NOP	No operation
OR s	Logical 'OR' of operand s and Acc.
OTDR	Load output port (C) with location (HL) decrement HL and B, repeat until B=0
OTIR	Load output port (C) with location (HL), increment HL, decrement B, repeat until B=0
OUT (C),r	Load output port (C) with Reg. r
OUT (n),A	Load output port (n) with Acc.
OUTD	Load output port (C) with location (HL), decrement HL and B
OUTI	Load output port (C) with location (HL), increment HL and decrement B
POP IX	Load IX with top of stack
POP IY	Load IY with top of stack

POP qq	Load Reg. pair qq with top of stack
PUSH IX	Load IX onto stack
PUSH IY	Load IY onto stack
PUSH qq	Load Reg. pair qq onto stack
RES b,m	Reset Bit b of operand m
RET	Return from subroutine
RET cc	Return from subroutine if condition cc is true
RETI	Return from interrupt
RETN	Return from non maskable interrupt
RL m	Rotate left through carry operand m
RLA	Rotate left Acc. through carry
RLC (HL)	Rotate location (HL) left circular
RLC (IX+d)	Rotate location (IX+d) left circclar
RLC (IY+d)	Rotate location (IY+d) left circular
RLC r	Rotate Reg. r left circular
RLCA	Rotate left circular Acc.
RLD	Rotate digit left and right between Acc. and location (HL).
RR m	Rotate right through carry operand m
RRA	Rotate right Acc. through carry
RRC m	Rotate operand m right circular
RRCA	Rotate right circular Acc.
RRD	Rotate digit right and left between Acc. and location (HL)
RST p	Restart to location p
SBC A,s	Subtract operand s from Acc. with carry
SBC HL,ss	Subtract Reg. pair ss from HL with carry
SCF	Set carry flag (C=1)
SET b,(HL)	Set Bit b of location (HL)
SET b,(IX+d)	Set Bit b of location (IX+d)
SET b,(IY+d)	Set Bit b of location (IY+d)
SET b,r	Set Bit b of Reg. r
SLA m	Shift operand m left arithmetic
SRA m	Shift operand m right arithmetic
SRL m	Shift operand m right logical
SUB s	Subtract operand s from Acc.
XOR s	Exclusive 'OR' operand s and Acc.

Appendix 3 ASCII table

Bit numbers									0	0	0	0	1	1	1	1
									0	0	1	1	0	0	1	1
									0	1	0	1	0	1	0	1
b_7	b_6	b_5	b_4	b_3	b_2	b_1	hex 1 hex 0		0	1	2	3	4	5	6	7
			0	0	0	0		0	NUL	DLE	SP	0	@	P	`	p
			0	0	0	1		1	SOH	DC1	!	1	A	Q	a	q
			0	0	1	0		2	STX	DC2	"	2	B	R	b	r
			0	0	1	1		3	ETX	DC3	#	3	C	S	c	s
			0	1	0	0		4	EOT	DC4	$	4	D	T	d	t
			0	1	0	1		5	ENQ	NAK	%	5	E	U	e	u
			0	1	1	0		6	ACK	SYN	&	6	F	V	f	v
			0	1	1	1		7	BEL	ETB	´	7	G	W	g	w
			1	0	0	0		8	BS	CAN	(	8	H	X	h	x
			1	0	0	1		9	HT	EM	)	9	I	Y	i	y
			1	0	1	0		10	LF	SUB	*	:	J	Z	j	z
			1	0	1	1		11	VT	ESC	+	;	K	[	k	{
			1	1	0	0		12	FF	FS	,	<	L	\	l	¦
			1	1	0	1		13	CR	GS	-	=	M	]	m	}
			1	1	1	0		14	SO	RS	.	>	N	^	n	~
			1	1	1	1		15	SI	US	/	?	O	¤	o	DEL

Answers to Questions

1.1 Yes.

1.2 Yes.

1.3 A low-level electrical signal.

1.4 To enable measurements to be made.

1.5 *(a)* To provide output power.
(b) To enable them to work.

1.6 By an algorithm.

1.7 A precise sequence of simpler tasks that unambiguously describe the process.

1.8 *(a)* System inputs.
(b) Internal conditions or states.

1.9 Convert energy and information from one physical form to another.

1.10 Examples include: resistance thermometer, floating-level gauge, record player pick-up and light meter.

1.11 Examples include: loudspeaker, motor, light and segmented display.

1.12 Examples include: oven, washing machine, radio, hi-fi and accounting system.

1.13 Refer back to examples in this chapter. In the case of an accounting system, the inputs are documents of financial transactions, e.g. invoices and cheque stubs. Outputs are records in books (or in computer) with totals, balance sheet summaries, etc.

1.14 Refer to Figure and text in Section 1.3.

1.15 Refer to Figure and text in Section 1.6.

1.16 Open dump-valve. Wait till empty.

1.17 Yes. No signal to say when drum is empty.

1.18 See Figure A overleaf (other arrangements possible).

1.19 *(a)* To show the sequence of steps.
(b) To briefly describe the purpose of each step.
(c) To define the algorithm, free of ambiguity.
(d) To make the description easy to understand.

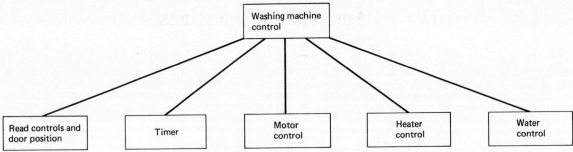

Figure A Hierarchy diagram for a washing machine controller (others possible)

1.20 To provide information.

1.21 A sensor is a measurement transducer. We expect to deduce the input conditions. This is why we use them.

1.22 An algorithm in a form that can be followed by a microelectronic device, or by a computer.

1.23 Processes inputs so as to change its internal state and its outputs, according to the algorithm that defines the process.

1.24 A microelectronic device that causes a predefined sequence of events.

1.25 The stored program.

2.1 No.

2.2 No, but the smaller the change the more difficult to observe it.

2.3 Possible quantities include weight, length, speed and volume.

2.4 To convert between analogue and digital quantities.

2.5 2.

2.6 +5V, the most positive of the two levels.

2.7 Yes. Without a source of power they will not work.

2.8 Because the outputs are logically related to the inputs.

2.9 To magnify the signal, i.e. to increase its size and power-driving capacity.

2.10 Analogue systems.

2.11 Because the shape of the graph of amplitude against time is a distinctive feature.

2.12 No.

2.13 Provides a signal path.

2.14 So that several sets of information can be sent over just one transmission path.

2.15 There should be eight combinations. Refer to Figure 2.14.

2.16 $2^8 = 256$.

2.17 8.

2.18 16.

2.19 So as to avoid ambiguous and therefore erroneous answers when they are changing.

2.20 Parallel needs many wires. Serial uses one wire, but is slower because data is transmitted bit by bit.

2.21 To select or point to the appropriate location.

2.22 $2^{10} = 1024$; usually called 1K.

2.23 Copying data from same place.

2.24 No. Like *load* it means copy.

2.25 A single bit of data.

2.26 The American Standard Code for Information Interchange. It is used to define binary codes to represent numbers and letters.

2.27 Gets the next instruction.

2.28 By binary numbers.

2.29 Suitable examples are:
(a) calculator, computer, digital watch;
(b) amplifier, radio, analogue watch.

2.30 No.

3.1 Resistors, capacitors and transistors.

3.2 A complete circuit within a single silicon chip.

3.3 No, they are made from silicon.

3.4 An integrated circuit.

3.5 Circuit components that are separate and individual.

3.6 Yes.

3.7 10,000 Ω.

3.8 100 picofarads.

3.9   A resistor.

3.10   See Figure 3.1.

3.11   To prevent voltages changing.

3.12   No. Passive components cannot amplify.

3.13   *(a)*   No.
       (b)   Yes.

3.14   Yes.

3.15   To connect components together.

3.16   To define the interconnection pattern.

3.17   A light-sensitive chemical used in the manufacture of printed circuits.

3.18   A rectangular piece of silicon containing a complete circuit.

3.19   Both use photographic techniques to define the required geometrical patterns.

3.20   A dual in-line package.

3.21   A diagram showing how the pins (of an integrated circuit) are allocated.

3.22   A transistor needs only three connections. Complete circuits need more.

3.23   SSI up to 10 elements
       MSI 10 to 100 elements
       LSI 100 to 1000 elements
       VLSI over 1000 elements.

4.1   No. It is only the processing unit.

4.2   Yes.

4.3   Microprocessor; memory; I/O unit; the bus.

4.4   The microprocessor controls the system and computes.
       The memory holds both program and data.
       I/O is the input/output arrangement.
       The bus enables the transfer of data between units.

4.5   Low-level electrical logic signals.

4.6   No, it is a copy operation.

4.7   *(a)*   Random access memory.
       (b)   Read only memory.
       (c)   Eraseable programmable ROM.

4.8   *(a)*   ROM and EPROM.

       (b)   All, ROM, RAM and EPROM.

       (c)   RAM.

4.9   ROM and EPROM.

4.10   To perform arithmetic and logic operations, i.e. to compute and manipulate data.

4.11   The microprocessor.

4.12   As binary words.

4.13   Input terminals, output terminals and any memory location.

4.14   By fetching the next instruction and looking at it.

4.15 $2^8 = 256$.

4.16   It must be executed.

4.17   Single-word memories within the processing unit.

4.18   A register. It is heavily used and usually holds the result of an operation.

4.19   In both cases it fetches the next instruction.

4.20   To identify the memory location to be accessed.

4.21   The control section of the processing unit.

4.22   The instruction register.

4.23   It depends. At least once to fetch the instruction. Perhaps again to read or write data but some instructions involve only data that was already within the processing unit.

4.24   The designer built-in logic circuits and truth tables so that it obeys the stored program.

4.25   To make the stored data quickly and easily available.

4.26   No, it is these features that account for the main differences between computer hardware.

4.27   Not so. Each type of computer has a particular set of instruction codes.

4.28 $2^{16} = 64K$ (here K means 1024) memory locations.

4.29   Refer to Figure 4.1 and Section 4.1.

5.1   To interface computers to the real world, i.e. to enable a computer to input and output information.

5.2   A visual display unit.

5.3 They both have a keyboard for data input.

5.4 Like a typewriter it prints on paper.

5.5 By punching a pattern of holes.

5.6 Yes.

5.7 Yes.

5.8 For reading yes, for writing no.

5.9 *(a)* and *(b)* magnetically coated plastic.

5.10 By using peripherals.

5.11 By means of a serial communication link.

5.12 A code to tell the computer which key(s) has been depressed.

5.13 *(a)* By messages from the computer.
 (b) Via the keyboard.
 (c) The computer is told which key has been depressed.
 (d) Only if programmed to echo.
 (e) A program that handles the peripheral.
 (f) It is the coding system normally used for the messages exchanged between the VDU and computer.

5.14 Printed paper produced by a computer.

5.15 *(a)* Messages to the computer.
 (b) Hard copy.
 (c) And perhaps paper tape as well.

5.16 *(a)* Key strokes.
 (b) Messages from the computer (or from own keyboard).
 (c) Paper tape.

5.17 To help the tape mechanism to move and position the tape.

5.18 Reading. Writing involves punching holes in the paper tape. It must be stopped and started for each row. Reading can be done whilst the tape is moving.

5.19 Yes. Each hole pattern can be seen by inspection. By comparing the binary pattern to a code chart each set of holes can be translated into a character. This can be done manually but the process is slow and tedious.

5.20 In both cases the data is stored as a magnetic pattern along a track.

5.21 Mag-tape
 Floppy disk.
 Paper tape unit.

5.22 To position it over the appropriate track.

5.23 Because they provide the computer with input information.

5.24 So that the computer can output music or synthetic speech.

5.25 To convert between analogue and digital signals.

5.26 See Figure 5.11 and the associated list.

5.27 Peripherals, in both cases.

5.28 For input: switches, buttons, knobs, joysticks.
For output: CRT screens, loudspeakers.

6.1 *(a) The system bus.* It includes:
Address bus – to allow the microprocessor to select storage and I/O locations.
Data bus – to allow data to be transferred in either direction. It is usually one byte wide.
Control bus – to allow the microprocessor to control memory and I/O.

(b) Memory In one block of memory the program is held. This is usually ROM. The data or any changeable program is held in RAM.

(c) I/O Ports The computer inputs and outputs via wires connected to the I/O ports. Each port is a byte wide, i.e. 8 bits. To the programmer the I/O port looks like a memory location.

(d) The microprocessor This is the heart of a microcomputer. It includes:
Control unit This unit controls everything.
Arithmetic and logic unit The ALU performs the data manipulation part of the relevant instructions.
Program counter and incrementer The program counter holds the instruction address. At the beginning of each new instruction it holds the address from which the new instruction is to be fetched. Before the instruction ends it holds the address of the next instruction.
Accumulator The accumulator is a special register. Data that is being worked on is often held in it. The results of data manipulation instructions are usually left in the accumulator.
Registers Registers are convenient places to hold data that is being worked on.
Flags Flags are single-bit memories that indicate the result of previous computation. They are used by conditional jump instructions.

6.2 The number held in the program counter.

6.3 Not enabled. A bar above always means the *opposite of*, a binary NOT.

6.4   To help provide a very precise clock of very high frequency (many megahertz).

6.5   The contents of the accumulator.

6.6   To make the computer start at a well defined storage location, usually zero.

6.7   See Figure 6.6.

6.8   An instruction that causes the program counter to be diverted from its normal progression.

6.9   A jump instruction causes the next instruction to be *out of normal sequence*. It causes the program to jump over a section of program. Say Jump to 85 was stored at location 25. Then after the execution of the instruction stored at 25, the next instruction would come from address 85. Instructions 26–84 would be jumped over.

6.10 The sequence of instructions would be, 60, 61, 62, 63, 62, 63, 62, 63, and so on.

6.11 The loop will continue for ever, unless terminated. It could be terminated by making the instruction at location 63 a conditional jump. It would need to be controlled by a flag that could be changed by the instruction at location 62.

6.12 An instruction is a binary word. A particular pattern represents a particular operation that the computer can perform.

6.13 An instruction set is the complete set of individual instructions that a particular microprocessor can execute.

6.14 Data movement
          Data manipulation
          Flow of control

6.15 Data movement: a, b,
          Data manipulation: c, d, e,
          Flow of control: f, g.

6.16 See Figure B opposite.

6.17 Refer to Figures 6.12–6.18 for guidance. It is very important that we can *visualise* what an instruction does.

Instruction	Op-Code								byte 2	byte 1
MOV A B	0	1	1	1	1	0	0	0		
MOV B A	0	1	0	0	0	1	1	1		
MVI A data	0	0	1	1	1	1	1	0	data	
MVI B data	0	0	0	0	0	1	1	0	data	
LDA addr	0	0	1	1	1	0	1	0		addr *
STA addr	0	0	1	1	0	0	1	0		addr *
ADD B	1	0	0	0	0	0	0	0		
ADI data	1	1	0	0	0	1	1	0	data	
INR A	0	0	1	1	1	1	0	0		
INR B	0	0	0	0	0	1	0	0		
DCR A	0	0	1	1	1	1	0	1		
DCR B	0	0	0	0	0	1	0	1		
CMP B	1	0	1	1	1	0	0	0		
CPI data	1	1	1	1	1	1	1	0	data	
STC	0	0	1	1	0	1	1	1		
JMP addr	1	1	0	0	0	0	1	1		addr *
JC	1	1	0	1	1	0	1	0		
JNC	1	1	0	1	0	0	1	0		
HLT	0	1	1	1	0	1	1	0		
NOP	0	0	0	0	0	0	0	0		

* low order byte first

Figure B Solution to Question 6.16

Index